Contemporary's

Number Power

a real world approach to math

3

Robert Mitchell

 Wright Group

Wright Group

Printed in the United States of America.

Send all inquiries to:
Wright Group/McGraw-Hill
130 E. Randolph, Suite 400
Chicago, IL 60681

ISBN 0-8092-2388-0

11 12 13 14 15 16 VHG 08 07 06

The McGraw-Hill Companies

Table of Contents

One-Step Equations

Multistep Equations

Special Equations

Graphing Equations

Inequalities

Polynomials

Factoring

To the Student

Welcome to *Algebra:*

Algebra is a powerful tool that makes solving complicated problems much easier. Algebra is the basic mathematical language of technical fields from auto mechanics to nursing and electronics. Also, algebra is a standard section on almost all educational and vocational tests, including GED, college entrance, civil service, and military entrance.

Number Power Algebra is designed to prepare you for taking a test and for pursuing further education or training. The book begins with a pretest that will give you an overview of the concepts you'll study in this book. The first part of the book, Building Number Power, provides step-by-step instruction in the fundamentals of algebra. This part is divided into 10 chapters. Each chapter ends with a review that you can use to check your understanding. At the end of Building Number Power, you can choose between two posttests. Posttest A asks you to solve problems and show your answers. Posttest B uses a multiple-choice format, much like many educational and vocational tests you may take later.

The second part of the book, Using Number Power, will give you a chance to apply algebraic skills in real-life situations. These applications are fun and will give you added chance to see the power of algebra.

Several skills that you may find helpful in your study of algebra are estimation, mental math, and calculator use. These skills are reviewed on pages 207 and 208. Also, for ready reference, a formula summary is provided on pages 209–210, and a glossary begins on page 211. The following icons will alert you to problems where estimation, mental math, or calculator use may be especially helpful.

 calculator icon

 estimation icon

 mental math icon

To get the most from your work, do each problem carefully. Check your answers with the answer key at the back of the book to make sure you are working accurately. Inside the back cover is a chart to help you keep track of your score on each exercise.

We hope you enjoy *Number Power Algebra!*

Algebra Pretest

This pretest will tell you which chapters of *Number Power Algebra* you need to work on and which you already have mastered. Do all the problems that you can. There is no time limit. Check your answers with the answer key at the back of the book. Then use the chart at the end of the test to find the pages where you need to work.

Write the addition shown by the number arrows.

1. _____

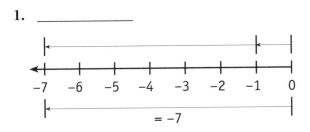

2. _____

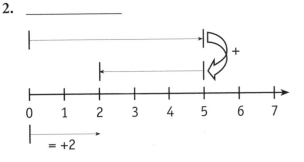

Subtract.

3. $8 - (-3) =$

4. $-12 - (-6) =$

Multiply or divide as indicated.

5. $(-6)(5) =$

6. $(-14) \div (-7) =$

Find the value of each power.

7. $4^2 =$

8. $(-3)^3 =$

Write each product or quotient as a single base and exponent.

9. $(6^4)(6^2) =$

10. $\dfrac{7^5}{7^2} =$

Find each square root below.

11. $\sqrt{64}$

12. $\sqrt{\dfrac{25}{49}} =$

13. Write the expression "3 times the quantity *n* plus 6" as an algebraic expression.

Find the value of each algebraic expression.

14. $8x - 4$ when $x = 3$

15. $(a + 2)(b - 2)$ when $a = 6, b = 4$

16. Use the formula $°C = \dfrac{5}{9} \, (°F - 32)$ to find the Celsius temperature (°C) when the Fahrenheit temperature (°F) is 212°F.

Circle the equation that is *not* represented by the drawing.

17.

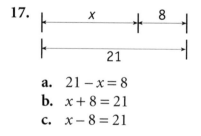

   a. $21 - x = 8$
   b. $x + 8 = 21$
   c. $x - 8 = 21$

For problems 18–22, solve each equation. Show all steps.

18. **a.** $x + 9 = 26$ **b.** $y - 5 = 17$

19. **a.** $3n = 18$ **b.** $\dfrac{n}{4} = 9$

20. **a.** $3x - 5 = 22$ **b.** $\dfrac{s}{4} + 8 = 20$

21. **a.** $5y + 3y = 64$ **b.** $4n - 7 = 2n + 9$

22. **a.** $2(x - 3) = 20$ **b.** $4(b + 2) = 2(b + 9)$

23. The ratio of two numbers is 4 to 3. The sum of the numbers is 49. What are the two numbers?

24. A recipe for 6 servings of dessert calls for 2 cups of flour. Using a proportion, determine how much flour is needed to make 16 servings.

25. Find the two correct values of x in the equation $3x^2 = y$ for $y = 75$.

26. Graph the line containing the points $(0, -4)$, $(2, 0)$, and $(3, 2)$.

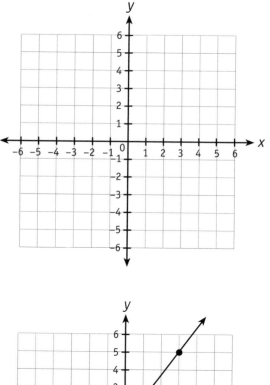

27. What is the slope of line M?

$$\text{slope} = \frac{\text{change in } y}{\text{change in } x}$$

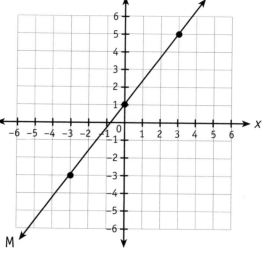

28. Graph the equation $y = 2x - 1$. As a first step, complete the Table of Values.

Table of Values

x	y
0	
1	
3	

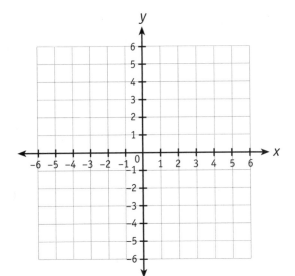

Write the inequality graphed on the number line.

29.

Values of n

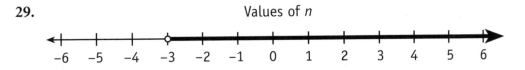

Solve the following inequalities.

30. $n + 4 \geq 6$

31. $3x - 5 < 16$

32. Graph the inequality $-3 \leq x < 5$.

Values of x

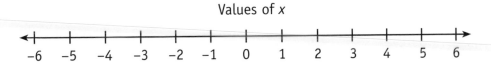

33. Circle the pair of like terms in the group of five terms below.

$-3x^2 \qquad 3y^2 \qquad 3x \qquad 5x^2 \qquad -2y^3$

Add or subtract as indicated.

34. $(5x^2 + 3x) + (-2x^2 - 4x) =$

35. $(3a^2 - 2b) - (2a^2 + 3b) =$

Multiply or divide as indicated.

36. $2n(3n^2 + 4n - 2) =$

37. $\dfrac{(14x^4 + 6x^3)}{2x^2} =$

Simplify each square root.

38. $\sqrt{75} =$

39. $\sqrt{36x^4 y} =$

Factor each expression as completely as possible.

40. $4x^2 + 10 =$

41. $12z^3 + 16z^2 =$

Simplify each quotient by dividing out the common factor.

42. $\dfrac{9r^2 + 12}{3} =$

43. $\dfrac{7c - 7d}{c - d} =$

Algebra Pretest Chart

This pretest can be used to identify algebra skills in which you are already proficient. If you miss only 1 question in a chapter, you may not need further study in that chapter. However, before you skip any lessons, complete the review at the end of that chapter. The end-of-chapter review will be a more precise indicator of your skill level.

PROBLEM NUMBERS	SKILL AREA	PRACTICE PAGES
1, 2, 3, 4	adding and subtracting signed numbers	12–17
5, 6	multiplying and dividing signed numbers	18–21
7, 8	evaluating a power	24–26
9, 10	simplifying a product or quotient	28–29
11, 12	finding a square root	32–33
13	writing algebraic expressions	38–39
14, 15	evaluating algebraic expressions	44–46
16	evaluating formulas	47–48
17	picturing an equation	52–53
18	solving addition and subtraction equations	56–58
19	solving multiplication and division equations	59–61
20	solving multistep equations	68–71
21	solving an equation with separated unknowns	72–73
22	solving an equation with parentheses	78–79
23, 24	using ratios and proportions	86–91
25	solving a quadratic equation	96–97
26	graphing a line	103
27	finding the slope of a line	104–105
28	graphing a linear equation	106–107
29, 32	graphing an inequality	117–118, 123–125
30, 31	solving an inequality	119–120
33	recognizing like terms	129
34, 35	adding and subtracting polynomials	130–133
36, 37	multiplying and dividing polynomials	134–138
38, 39	simplifying a square root	145
40, 41, 42, 43	factoring and dividing	146–149

Building
Number
Power

SIGNED NUMBERS

Introducing Signed Numbers

The study of algebra begins with **signed numbers**—the set of all positive and negative numbers. Look at the thermometer at the right. Both positive and negative numbers are shown.

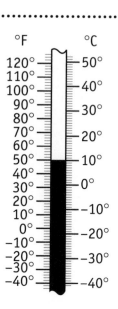

- **Positive numbers** are greater than zero. Positive temperatures are greater than zero degrees.
- **Negative numbers** are less than zero. Negative temperatures are less than zero degrees. Negative temperatures are preceded by a negative sign (−).

The numbers you are most familiar with are positive numbers. Negative numbers, though not as common as positive numbers, also have many uses.

- 15° below 0° is written as −15°.

- A $\frac{1}{2}$ dollar drop in a stock market price is written $-\frac{1}{2}$.

- 35 feet below sea level is written as −35 feet.

Zero (0) separates positive numbers from negative numbers. Zero is neither positive nor negative.

Positive numbers are written with a positive sign (+) or with no sign at all. Negative numbers are always written with a negative sign (−).

Positive 20 is written as +20 or 20.   Negative 20 is written as −20.

Positive $5\frac{3}{4}$ is written as $+5\frac{3}{4}$ or $5\frac{3}{4}$.   Negative 7.5 is written as −7.5.

Zero has no sign and is always written as 0.

(**Remember:** A number without a sign is always a positive number.)

Write each number as a signed number with a + sign or a − sign.

1. positive 8 =   positive 3 =   positive 12 =   positive 7 =

2. negative 5 =   negative 4 =   negative 19 =   negative 8 =

3. positive $2\frac{1}{2}$ =   positive 4.75 =   negative $5\frac{2}{3}$ =   negative $\frac{1}{4}$ =

Identify each number as either positive or negative.

4. +8   −7 $\frac{3}{4}$ −6

5. $2\frac{2}{3}$ $+4\frac{1}{2}$ −9.3 $-\frac{3}{8}$

The Number Line

Signed numbers can be pictured on a **number line.** The value of numbers increases as you move from left to right.

Positive numbers are written to the right of 0.
- Every positive number is greater than 0.
- Every positive number is greater than any negative number.

Negative numbers are written to the left of 0.
- Every negative number is less than 0.
- Every negative number is less than any positive number.
- You can read a negative number in any of three ways.

> **Did You Know . . . ?**
> Whole numbers, fractions, and decimals can all be represented on a number line.

EXAMPLE −4 can be read as *negative 4, minus 4,* or *4 below zero.*

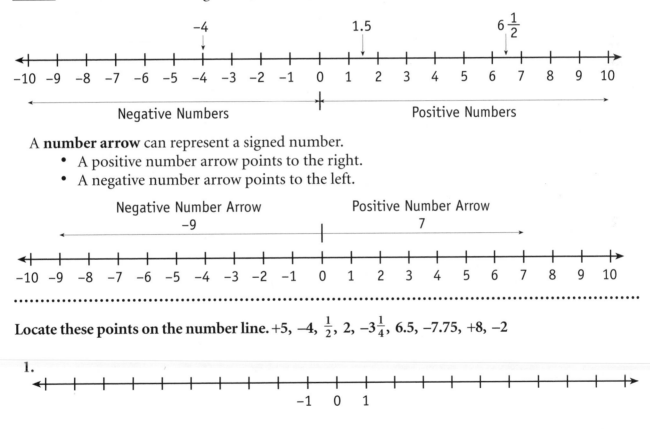

A **number arrow** can represent a signed number.
- A positive number arrow points to the right.
- A negative number arrow points to the left.

Locate these points on the number line. +5, −4, $\frac{1}{2}$, 2, $-3\frac{1}{4}$, 6.5, −7.75, +8, −2

1.

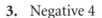

Draw a number arrow for each signed number.

2. Positive 3

3. Negative 4

Adding Signed Numbers

In arithmetic, you learn to add and subtract positive numbers. In algebra, you learn that you can also add and subtract negative numbers. To avoid confusion, a signed number may be enclosed in parentheses. Look at these examples.

Sum or Difference	Meaning	Simplified Way of Writing
$+5 + (+6)$	positive 5 plus positive 6	$5 + 6$
$+3 + (-4)$	positive 3 plus negative 4	$3 + (-4)$
$-2 + (-3)$	negative 2 plus negative 3	$-2 + (-3)$

(**Note:** A positive sign indicates either addition *or* a positive number. A negative sign indicates either subtraction *or* a negative number.)

You will use number arrows to help learn the rules for adding signed numbers.

> **Rule for adding numbers that have the same sign:**
> Add the numbers and give the sum the same sign as the numbers.

Example 1, though simple, shows the use of number arrows.

<u>EXAMPLE 1</u> Add +3 and +2.

STEP 1 Add the numbers. $3 + 2 = 5$

STEP 2 Give the sum a positive sign.

$+3$ and $+2 = \mathbf{+5}$

or $3 + 2 = \mathbf{5}$

ANSWER: +5 or 5

Adding Number Arrows

Add a +3 arrow and a +2 arrow.

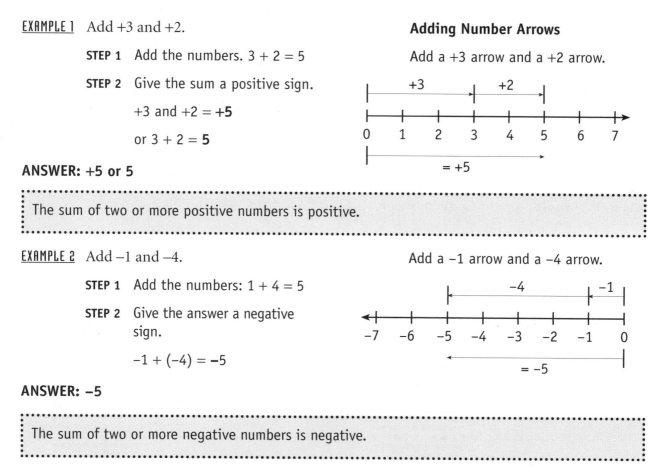

> The sum of two or more positive numbers is positive.

<u>EXAMPLE 2</u> Add −1 and −4.

STEP 1 Add the numbers: $1 + 4 = 5$

STEP 2 Give the answer a negative sign.

$-1 + (-4) = \mathbf{-5}$

ANSWER: −5

Add a −1 arrow and a −4 arrow.

> The sum of two or more negative numbers is negative.

Write the addition shown by the number arrows. The first problem is done as an example.

1. $4 + 3 = 7$

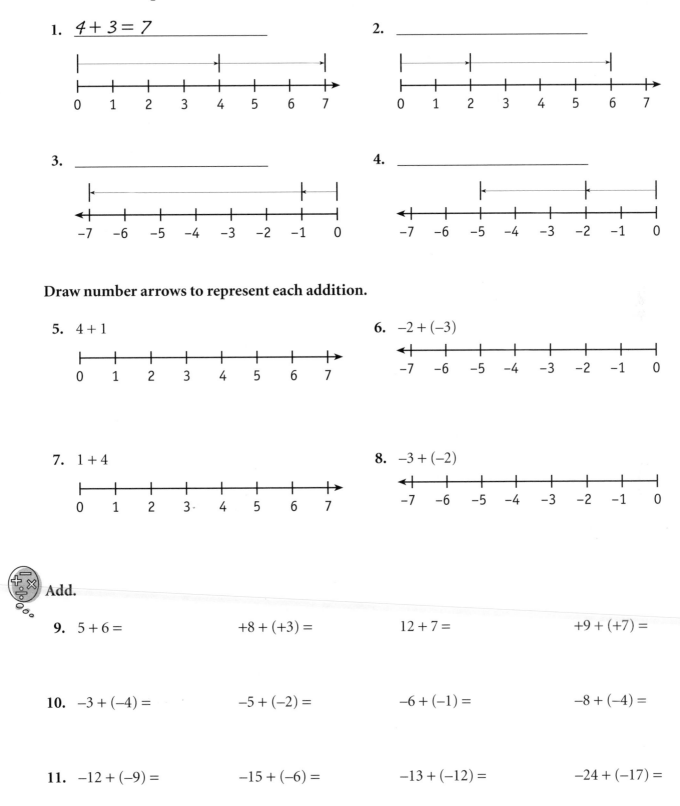

2. _____

3. _____

4. _____

Draw number arrows to represent each addition.

5. $4 + 1$

6. $-2 + (-3)$

7. $1 + 4$

8. $-3 + (-2)$

Add.

9. $5 + 6 =$ $+8 + (+3) =$ $12 + 7 =$ $+9 + (+7) =$

10. $-3 + (-4) =$ $-5 + (-2) =$ $-6 + (-1) =$ $-8 + (-4) =$

11. $-12 + (-9) =$ $-15 + (-6) =$ $-13 + (-12) =$ $-24 + (-17) =$

> **Rule for adding a positive number and a negative number:**
> Find the difference between the two numbers and give the sum the sign of the greater number.

The sum of a positive number and a negative number may be positive, negative, or zero.

<u>EXAMPLE 3</u> Add +5 and −2.

    STEP 1 Find the difference between 5 and 2.

             5 − 2 = 3

    STEP 2 Give the sum a positive sign (5 is greater than 2).

Add a +5 arrow and a −2 arrow.

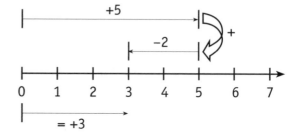

ANSWER: +3

(**Note:** Adding a negative number is the same as subtracting a positive number. $5 + (−2) = 5 − (+2) = 3$)

<u>EXAMPLE 4</u> Add +2 and −4.

    STEP 1 Find the difference between 4 and 2.

             4 − 2 = 2

    STEP 2 Give the answer a negative sign (4 is greater than 2).

Add a +2 arrow and a −4 arrow.

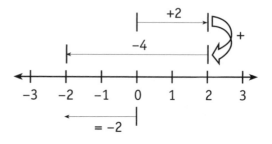

ANSWER: −2

<u>EXAMPLE 5</u> Add −3 and +3.

    Find the difference between 3 and 3.

    3 − 3 = 0

Add a −3 arrow and a +3 arrow.

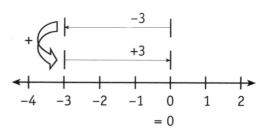

ANSWER: 0

> Two numbers that differ only by sign are called **opposites**. The sum of opposites is always 0.

Write the addition shown by the number arrows.

12. _____

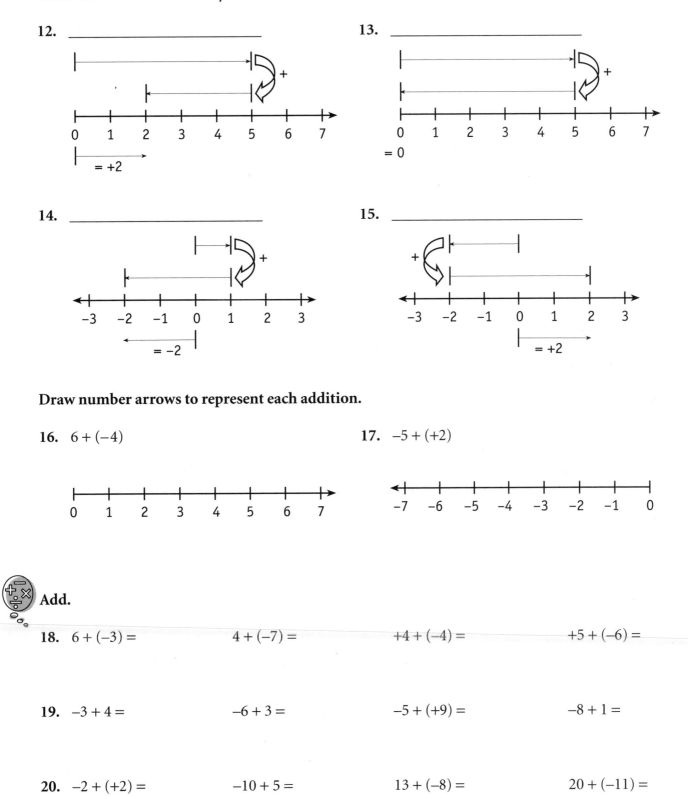

= +2

13. _____

= 0

14. _____

= −2

15. _____

= +2

Draw number arrows to represent each addition.

16. 6 + (−4)

17. −5 + (+2)

Add.

18. 6 + (−3) =   4 + (−7) =   +4 + (−4) =   +5 + (−6) =

19. −3 + 4 =   −6 + 3 =   −5 + (+9) =   −8 + 1 =

20. −2 + (+2) =   −10 + 5 =   13 + (−8) =   20 + (−11) =

Subtracting Signed Numbers

> **Rule for subtracting signed numbers:**
> Change the sign of the number being subtracted, and then add the numbers.

EXAMPLE 1 Subtract +5 from +8.

> **STEP 1** Write +8 – (+5)
>
> or 8 – 5.
>
> **STEP 2** Change the sign of +5 to –5, and then add.
>
> 8 – 5 = 8 + (–5) = **3**

In algebra, subtracting a signed number is the same as adding the same number with its sign reversed.

ANSWER: 3

(**Note:** You should continue to subtract positive numbers in the usual way! Example 1 just shows you a new way to think about subtraction.)

EXAMPLE 2 Subtract 12 from 3.

> **STEP 1** Write 3 – 12.
>
> **STEP 2** Change the sign of 12 to –12 and then add.
>
> 3 – 12 = 3 + (–12) = **–9**

In algebra, you can subtract a greater number from a lesser number! Signed numbers make this possible.

ANSWER: –9

EXAMPLE 3 Subtract –4 from 7.

> **STEP 1** Write 7 – (–4).
>
> **STEP 2** Change the sign of –4 to +4 and then add.
>
> 7 – (–4) = 7 + 4 = **11**

Subtracting signed numbers can be thought of as finding the distance between two points on a number line.

ANSWER: 11

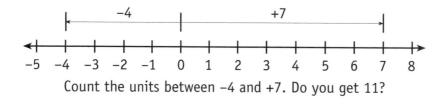

Count the units between –4 and +7. Do you get 11?

Circle the subtraction that is represented on each number line.

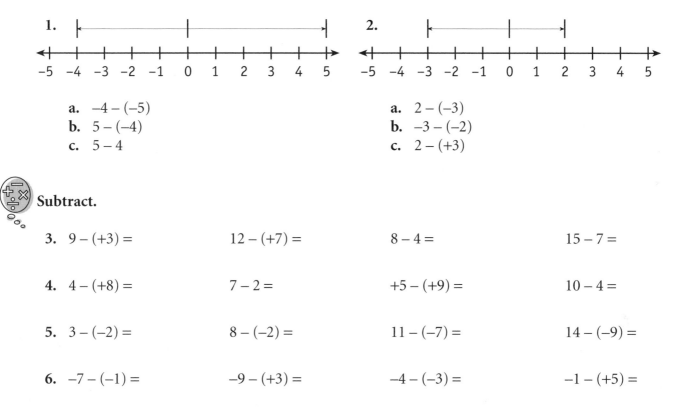

1.

    a. $-4 - (-5)$
    b. $5 - (-4)$
    c. $5 - 4$

2.

    a. $2 - (-3)$
    b. $-3 - (-2)$
    c. $2 - (+3)$

Subtract.

3. $9 - (+3) =$ $12 - (+7) =$ $8 - 4 =$ $15 - 7 =$

4. $4 - (+8) =$ $7 - 2 =$ $+5 - (+9) =$ $10 - 4 =$

5. $3 - (-2) =$ $8 - (-2) =$ $11 - (-7) =$ $14 - (-9) =$

6. $-7 - (-1) =$ $-9 - (+3) =$ $-4 - (-3) =$ $-1 - (+5) =$

Solve.

7. At 9:00 P.M. the temperature in Chicago was –6°F. By morning the temperature was +7°F. How many degrees did the temperature rise overnight?

8. On January 23, 1971, the temperature in Prospect Creek, Alaska, fell to –79°F. How many degrees below freezing (32°F) did the temperature fall?

9. Mount McKinley in Alaska, the highest mountain in North America, is 20,320 feet above sea level. Death Valley, the lowest point, is 280 feet below sea level. What is the difference in height between Mount McKinley and Death Valley?

10. Mount Whitney in California, the highest mountain in the U.S., excluding Alaska, is 14,495 feet high. What is the difference in height between Mount Whitney and Death Valley? (See problem 9.)

Multiplying Signed Numbers

In algebra, as well as arithmetic, multiplication can be shown in three ways.
- by a times sign: 5×4
- by a dot: $5 \cdot 4$
- by parentheses: $5(4)$

Rules for multiplying signed numbers:
- If the signs of the numbers are alike, multiply the numbers and give the product a positive sign.

  positive × positive = positive   negative × negative = positive

- If the signs of the numbers are different, multiply the numbers and give the product a negative sign.

  positive × negative = negative   negative × positive = negative

---

EXAMPLE 1   Multiply +4 times +3.

    STEP 1   Multiply the numbers. $4 \times 3 = 12$

    STEP 2   Because the signs are both positive, the product is positive.

Multiply: $+4 \times +3 = \mathbf{+12}$

EXAMPLE 2   Find the product of −5 times −6.

    STEP 1   Multiply the numbers. $5(6) = 30$

    STEP 2   Because the signs are both negative, the product is positive.

Multiply: $-5(-6) = \mathbf{+30}$

EXAMPLE 3   Multiply +2 times −7.

    STEP 1   Multiply the numbers. $2(7) = 14$

    STEP 2   Because the signs are different (one positive, one negative), the product is negative.

Multiply: $+2(-7) = \mathbf{-14}$

EXAMPLE 4   Find the product of −9 times +6.

    STEP 1   Multiply the numbers. $9 \cdot 6 = 54$

    STEP 2   Because the signs are different (one negative, one positive), the product is negative.

Multiply: $-9 \cdot +6 = \mathbf{-54}$

Multiply. (Remember: Like signs give positive products. Unlike signs give negative products.)

1. $3 \cdot 6 =$ $+7 \cdot -5 =$ $+8 \cdot +4 =$ $-6 \cdot -9 =$

2. $-8 \times -7 =$ $5 \times 4 =$ $6 \times -8 =$ $-3 \times 4 =$

3. $(+6)(+6) =$ $(-4)(-7) =$ $7(6) =$ $(-2)(-9) =$

4. $(8)(-5) =$ $(-3)(+3) =$ $(9)(-4) =$ $(-5)(6) =$

5. $(-9)(-5) =$ $(-9)(+6) =$ $(-7)(-3) =$ $(+6)(-4) =$

When you multiply more than two signed numbers, multiply them two at a time or use the shortcut given by the following rule.

Rules for multiplying more than two signed numbers:
Multiply all the numbers together.
- If there are an even number of negative signs, give the product a positive sign.
- If there are an odd number of negative signs, give the product a negative sign.

Multiply. Two problems are done as examples.

6. $(-5)(+2)(-3) = +30$ $(-2)(-2)(+4) =$ $(-2)(+3)(-2)(+4) =$
two negative signs

7. $(-3)(-4)(-2) = -24$ $(-1)(-3)(-6) =$ $(-1)(-3)(-2)(-2)(-3) =$
three negative signs

8. $-3 \cdot -4 =$ $-5 \cdot -2 \cdot -4 =$ $-1 \cdot -4 \cdot -6 \cdot -2 =$

9. $-2 \times -4 \times -2 =$ $-3 \times 5 \times -2 =$ $-1 \times -4 \times 2 =$

Dividing Signed Numbers

Division can also be shown in three ways.

- by a division sign: $15 \div 3$
- by a slash: $15/3$
- by a bar: $\frac{15}{3}$

Rules for dividing signed numbers:

- If the signs of the numbers are alike, divide the numbers and give the quotient a positive sign.

  positive ÷ positive = positive   negative ÷ negative = positive

- If the signs of the numbers are different, divide the numbers and give the quotient a negative sign.

  positive ÷ negative = negative   negative ÷ positive = negative

---

EXAMPLE 1   Divide +24 by +6.

   STEP 1   Divide the numbers. $24 \div 6 = 4$

   STEP 2   Because the signs are both positive, the quotient is positive.

Divide: $+24 \div +6 = \mathbf{+4}$

EXAMPLE 2   Find the quotient of −25 divided by −5.

   STEP 1   Divide the numbers. $\frac{25}{5} = 5$

   STEP 2   Because the signs are both negative, the quotient is positive.

Divide: $\frac{-25}{-5} = \mathbf{+5}$

EXAMPLE 3   Divide +30 by −10.

   STEP 1   Divide the numbers. $30 \div 10 = 3$

   STEP 2   Because the signs are different (one positive, one negative), the quotient is negative.

Divide: $+30 \div -10 = \mathbf{-3}$

EXAMPLE 4   Find the quotient of −16 divided by +4.

   STEP 1   Divide the numbers. $\frac{16}{4} = 4$

   STEP 2   Because the signs are different (one negative, one positive), the quotient is negative.

Divide: $\frac{-16}{+4} = \mathbf{-4}$

Divide. (Remember: Like signs give positive quotients. Unlike signs give negative quotients.)

1. $\dfrac{-24}{+4} =$ $\dfrac{+30}{-6} =$ $\dfrac{-42}{-7} =$ $\dfrac{+63}{+9} =$

2. $\dfrac{-104}{-8} =$ $\dfrac{-68}{2} =$ $\dfrac{54}{3} =$ $\dfrac{70}{-5} =$

3. $-18 \div 6 =$ $72 \div -8 =$ $81 \div 9 =$ $-48 \div -6 =$

4. $100 \div -4 =$ $-28 \div 2 =$ $-150 \div -5 =$ $96 \div -4 =$

5. $12/-6 =$ $-4/2 =$ $-18/-3 =$ $-28/7 =$

6. $42/-3 =$ $-60/4 =$ $-72/-6 =$ $-125/5 =$

> Sometimes, the answer to a division problem is a fraction. The fraction should be reduced to lowest terms and the sign placed in front of the fraction.

Divide. The first problem in each row is done as an example.

7. $-6 \div 9 = \dfrac{-6}{9} = -\dfrac{2}{3}$ $-9 \div 12 =$ $-5 \div 10 =$ $-8 \div 20 =$

8. $-3 \div -6 = \dfrac{-3}{-6} = \dfrac{1}{2}$ $-4 \div -10 =$ $-12 \div -16 =$ $-8 \div -12 =$

9. $\dfrac{5}{-25} = -\dfrac{5}{25} = -\dfrac{1}{5}$ $\dfrac{-50}{125} =$ $\dfrac{35}{-100} =$ $\dfrac{-100}{150} =$

Signed Numbers Review

Solve the problems below. When you finish, check your answers at the back of the book. Then correct any errors.

Write each number as a signed number with a + sign or a − sign.

1. positive 13 =   positive $2\frac{1}{2}$ =   negative 26 =   negative $\frac{7}{8}$ =

Locate these points on the number line. $-10,\ -7,\ -4\frac{1}{2},\ -2,\ -\frac{1}{4},\ 1\frac{1}{2},\ 5,\ 8,\ 9\frac{3}{4}$

2.

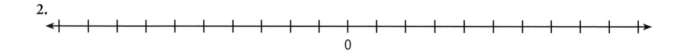

Write the addition shown by each pair of number arrows.

3. _____   4. _____

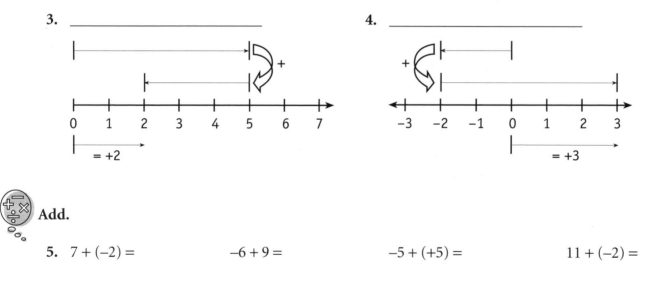

Add.

5. $7 + (-2) =$ $-6 + 9 =$ $-5 + (+5) =$ $11 + (-2) =$

6. $-8 + (-4) =$ $-9 + (-3) =$ $-6 + (+9) =$ $-7 + 3 =$

Subtract.

7. $8 - (+3) =$ $6 - (+6) =$ $5 - (+8) =$ $12 - (-7) =$

8. $-5 - (-7) =$ $-11 - 2 =$ $-5 - (+3) =$ $-9 - (-3) =$

Solve.

9. At 10:00 A.M. the temperature in New York City was –6°F. By 3:00 P.M. the temperature rose to +17°F. How many degrees did the temperature rise?

Multiply.

10. $(9)(-3) =$ $(-4)(+7) =$ $(6)(-3) =$ $(-8)(2) =$

11. $(-6)(-5) =$ $(+5)(+3) =$ $(-8)(-4) =$ $(9)(6) =$

12. $(-2)(-4)(-3) =$ $(-2)(+3)(-6) =$ $(-1)(-3)(-2)(-3) =$

Divide.

13. $\dfrac{-30}{+6} =$ $\dfrac{+28}{-4} =$ $\dfrac{-48}{-6} =$ $\dfrac{+56}{+7} =$

14. $24 \div -6 =$ $-64 \div -8 =$ $49 \div -7 =$ $-54 \div 9 =$

15. $18/-3 =$ $-14/7 =$ $-24/-8 =$ $-27/3 =$

16. $-14 \div -16 =$ $8 \div -10 =$ $-6/8 =$ $-10/-15 =$

POWERS AND ROOTS

What Is a Power?

A **power** is the product of a number multiplied by itself one or more times. Four to the third power means $4 \times 4 \times 4$.

A power is commonly written as a **base** and an **exponent**.

$4 \times 4 \times 4$ is written $4^3 \leftarrow$ exponent
$\uparrow$
base

The 4 is called the *base*. The base is the number being multiplied.

The 3 is called the *exponent*. The exponent tells how many times the base is written in the product.

Look at these examples.

Product	As a base and exponent	In words
$3 \cdot 3$	3^2	3 to the second power *or* 3 squared
$(-4)(-4)(-4)$	$(-4)^3$	negative 4 to the third power *or* negative 4 cubed
$\frac{1}{2} \times \frac{1}{2} \times \frac{1}{2} \times \frac{1}{2}$	$\left(\frac{1}{2}\right)^4$	$\frac{1}{2}$ to the fourth power

(**Note:** A number raised to the second power is often called *squared* and raised to the third power is often called *cubed*. These are the only two powers that have special names.)

The **value** of a power is found by multiplication.

__EXAMPLE 1__ Find the value of 4^4.

    STEP 1 Write the power 4^4 as a product. $4^4 = (4)(4)(4)(4)$

    STEP 2 Multiply the first two terms together.
             $(4)(4) = 16$ $= (16)(4)(4)$

    STEP 3 Multiply the answer in Step 2 times the
             next number. $(16)(4) = 64$ $= (64)(4)$

    STEP 4 Continue multiplying until you have used
             all the terms. $(64)(4) = 256$ **= 256**

ANSWER: 256

<u>EXAMPLE 2</u> Find the value of $\left(-\frac{2}{3}\right)^3$.

  STEP 1 Write the power $\left(-\frac{2}{3}\right)^3$ as a product.

  STEP 2 Multiply the first two terms. $\left(-\frac{2}{3}\right)\left(-\frac{2}{3}\right) = +\frac{4}{9}$

  STEP 3 Multiply the remaining terms.

$$\left(-\frac{2}{3}\right)^3 = \left(-\frac{2}{3}\right)\left(-\frac{2}{3}\right)\left(-\frac{2}{3}\right)$$
$$= \left(+\frac{4}{9}\right)\left(-\frac{2}{3}\right)$$
$$= -\frac{8}{27}$$

ANSWER: $-\frac{8}{27}$

  Be sure that the sign of your answer is correct.

....

Complete the following chart.

	Product	Base/Exponent	In Words
1.	$(-1)(-1)$	$(-1)^2$	−1 to the second power or −1 squared
2.	$(+5)(+5)$		
3.	$\left(+\frac{1}{2}\right)\left(+\frac{1}{2}\right)$		
4.	6×6		
5.	$7 \cdot 7 \cdot 7$		7 to the third power or 7 cubed
6.	$(-3)(-3)(-3)$		
7.	$(6)(6)(6)$		
8.	$\left(-\frac{3}{4}\right)\left(-\frac{3}{4}\right)\left(-\frac{3}{4}\right)$		
9.	$2 \times 2 \times 2 \times 2$		2 to the fourth power
10.	$5 \cdot 5 \cdot 5 \cdot 5$		
11.	$(-4)(-4)(-4)(-4)$		
12.	$\left(\frac{1}{4}\right)\left(\frac{1}{4}\right)\left(\frac{1}{4}\right)\left(\frac{1}{4}\right)$		

Find the value of each power.

13. $5^2 =$ $(-2)^2 =$ $(9)^2 =$ $(-6)^2 =$

14. $5^3 =$ $(3)^4 =$ $(-10)^3 =$ $(-3)^4 =$

15. $\left(\frac{2}{3}\right)^2 =$ $\left(-\frac{4}{5}\right)^3 =$ $\left(\frac{1}{2}\right)^4 =$ $\left(\frac{3}{4}\right)^3 =$

There are two special cases of powers.

> **Rule:** Any number to the first power is that number.

EXAMPLE 3 $8^1 = 8$

An exponent of 1 means the base 8 is written only once and is not multiplied.

> **Rule:** Any number to the 0 power is 1.

EXAMPLE 4 $6^0 = 1$

This is the hardest rule to remember. A zero exponent means *a number divided by itself*.

To simplify an expression containing more than one power, find the value for each power and then follow the rules for adding and subtracting signed numbers.

EXAMPLE 5   Simplify the expression $(-4)^3 - 3^4 + 7^0$.

    STEP 1   Find the value of each power.

        $(-4)^3 = -64$ $3^4 = 81$ $7^0 = 1$

    STEP 2   Substitute the values and follow the rules for adding and subtracting signed numbers.

$$(-4)^3 - 3^4 + 7^0 = -64 - 81 + 1$$
$$= -64 + (-81) + 1 \text{ (Adding } -81 \text{ is the same as subtracting 81.)}$$
$$= -145 + 1$$
$$= \mathbf{-144}$$

ANSWER: −144

Find the value of each expression.

16. $8^2 - 3^1 =$ $9^2 - 2^0 =$ $7^0 + 4^1 =$

17. $5^0 + (-3)^2 + 4^1 =$ $4^3 - 3^2 + 5^0 =$ $(-2)^3 - (-3)^3 + 2^1 =$

18. $\left(\frac{1}{2}\right)^2 + \left(\frac{1}{2}\right)^1 =$ $\left(-\frac{3}{4}\right)^2 - \left(\frac{1}{2}\right)^1 =$ $\left(\frac{3}{4}\right)^2 + \left(-\frac{1}{2}\right)^1 + \left(\frac{1}{2}\right)^0 =$

Multiplication and Division of Powers

You may need to find the value of a product or quotient of powers. As your first step, find the value of each power and then multiply or divide as indicated.

EXAMPLE 1 Find the value of the product $3^2 \cdot 2^4$.

STEP 1 Find the value of each power.

$3^2 = 9$ and $2^4 = 16$

STEP 2 Multiply the values found in Step 1.

$3^2 \cdot 2^4 = 9 \cdot 16 = \mathbf{144}$

ANSWER: 144

EXAMPLE 2 Find the value of the quotient $\dfrac{4^3}{2^2}$.

STEP 1 Find the value of each power.

$4^3 = 64$ and $2^2 = 4$

STEP 2 Divide the values found in Step 1.

$\dfrac{4^3}{2^2} = \dfrac{64}{4} = \mathbf{16}$

ANSWER: 16

Find the value of each product or quotient. The first problem in each row is done as an example.

1. $2^3 \cdot 4^2 = 8 \cdot 16$
    $= 128$

    $4^2 \cdot 3^3 =$

    $(-2)^3 \cdot 5^2 =$

2. $\left(\frac{1}{2}\right)^3 \cdot 5^2 = \left(\frac{1}{8}\right) \cdot 25$
    $= 3\frac{1}{8}$

    $\left(\frac{1}{5}\right)^2 \cdot 5^3 =$

    $\left(\frac{1}{4}\right)^2 \cdot \left(\frac{4}{3}\right)^3 =$

3. $\dfrac{5^3}{7^2} = \dfrac{125}{49}$
    $= 2\frac{27}{49}$

    $\dfrac{3^4}{7^2} =$

    $\dfrac{(-4^3)}{2^4} =$

To solve more complicated problems, first find the value of each power. Then use cancellation when possible to simplify the multiplication and division.

4. $\dfrac{8^2 \cdot 4^1}{2^3 \cdot 3^2} = \dfrac{\overset{8}{64} \cdot 4}{8_1 \cdot 9}$
    $= \dfrac{8 \cdot 4}{9}$
    $= \dfrac{32}{9}$
    $= 3\frac{5}{9}$

    $\dfrac{7^2 \cdot 2^3}{4^2 \cdot 3^2} =$

    $\dfrac{4^2 \cdot 5^1}{2^3 \cdot (-6)^2} =$

Simplifying Products and Quotients with Like Bases

Products of Powers with Like Bases

Multiplication often involves **like bases** (same base number). As the examples show, a product of powers with like bases can be simplified by adding exponents.

EXAMPLE 1 Simplify $3^4 \times 3^2$.

One way to simplify this product is to write each power as a product and then count 3s.

$3^4 \times 3^2 = 3 \times 3 \times 3 \times 3 \times 3 \times 3 = \mathbf{3^6}$ (There are six 3s in the product.)

ANSWER: 3^6

A shortcut is to add the exponents of each power.

$3^4 \times 3^2 = 3^{4+2} = \mathbf{3^6}$

Seeing this, you can write the rule for simplifying a product of powers with like bases.

Rule for simplifying a product of powers with like bases:
To multiply numbers with the same base, add the exponents.

EXAMPLE 2 Simplify $2^2 \cdot 2^3$.

$2^2 \cdot 2^3 = 2^{2+3} = \mathbf{2^5}$

ANSWER: 2^5

EXAMPLE 3 Simplify $(4^3)(4^2)(4^1)$.

$(4^3)(4^2)(4^1) = 4^{3+2+1} = \mathbf{4^6}$

ANSWER: 4^6

Add exponents to simplify each product of powers. The first problem in each row is done as an example.

1. $2^3 \times 2^4 = \mathbf{2^{3+4}} = \mathbf{2^7}$ $3^2 \times 3^2 =$ $4^4 \times 4^3 =$

2. $5^2 \cdot 5^1 = \mathbf{5^{2+1}} = \mathbf{5^3}$ $7^3 \cdot 7^1 =$ $4^3 \cdot 4^1 =$

3. $(6^2)\,(6^1)\,(6^0) = \mathbf{6^{2+1+0}}$ $(2^2)(2^3)(2^0) =$ $(8^3)(8^2)(8^2) =$
    $ = \mathbf{6^3}$

4. $4^3 \cdot 4^2 \cdot 3^4 = \mathbf{4^{3+2}} \cdot \mathbf{3^4}$ $(5^7)(5^5)(2^3) =$ $7^4 \times 5^3 \times 7^2 =$
    $ = \mathbf{4^5 \cdot 3^4}$
    (4 and 3 are not like bases)

Quotients of Powers with Like Bases

Division also may involve like bases. A quotient of powers with like bases can be simplified by subtracting exponents.

EXAMPLE 4 Simplify the quotient $4^7 \div 4^5$.

The long way is to write each power as a product and then cancel 4s.

$\dfrac{4^7}{4^5} = \dfrac{4 \times 4 \times 4 \times 4 \times 4 \times 4 \times 4}{4 \times 4 \times 4 \times 4 \times 4} = 4 \times 4 = \mathbf{4^2}$ (There are two 4s left in the product.)

ANSWER: 4^2

A shortcut is to subtract the exponents of each power.

$\dfrac{4^7}{4^5} = 4^{7-5} = \mathbf{4^2}$

Seeing this, you can write the rule for simplifying a quotient of powers with like bases.

> **Rule for simplifying a quotient of powers with like bases:**
> To divide numbers with the same base, subtract the exponents.

EXAMPLE 5 Simplify $2^6 \div 2^3$.

$2^6 \div 2^3 = 2^{6-3} = \mathbf{2^3}$

ANSWER: 2^3

EXAMPLE 6 Simplify $7^4 \div 7^3$.

$7^4 \div 7^3 = 7^{4-3} = \mathbf{7^1 = 7}$

ANSWER: 7

Subtract exponents to simplify each quotient of powers. The first problem in each row is done as an example.

5. $5^3 \div 5^1 = \boldsymbol{5^{3-1} = 5^2}$ $\qquad$ $2^6 \div 2^2 =$ $\qquad$ $8^3 \div 8^1 =$

6. $7^3 \div 7^2 = \boldsymbol{7^{3-2} = 7^1 = 7}$ $\qquad$ $8^4 \div 8^3 =$ $\qquad$ $5^7 \div 5^6 =$

7. $\dfrac{4^7}{4^5} = \boldsymbol{4^{7-5} = 4^2}$ $\qquad$ $\dfrac{8^6}{8^3} =$ $\qquad$ $\dfrac{7^6}{7^5} =$

8. $\dfrac{6^5 \cdot 4^2}{6^2} = \boldsymbol{6^{5-2} \cdot 4^2}$ $\qquad$ $\dfrac{10^9 \cdot 5^4}{10^7} =$ $\qquad$ $\dfrac{9^5 \cdot 7^6}{9^3 \cdot 7^3} =$

$\boldsymbol{= 6^3 \cdot 4^2}$

(6 and 4 are not like bases)

Negative Exponents

Negative numbers are also used as exponents. A **negative exponent** stands for the number found by inverting a power. For example, $4^{-2} = \frac{1}{4^2} = \frac{1}{4 \times 4}$. Below are more examples.

Power with Negative Exponent	Meaning	In Words	Value
3^{-1}	$\frac{1}{3}$	3 to the minus 1st power	$\frac{1}{3}$
5^{-2}	$\frac{1}{5 \times 5}$	5 to the minus 2nd power	$\frac{1}{25}$
4^{-3}	$\frac{1}{4 \times 4 \times 4}$	4 to the minus 3rd power	$\frac{1}{64}$
2^{-4}	$\frac{1}{2 \times 2 \times 2 \times 2}$	2 to the minus 4th power	$\frac{1}{16}$

The rules that apply to positive exponents also apply to negative exponents.

Add exponents when multiplying like bases.

Subtract exponents when dividing like bases.

EXAMPLE 1 Simplify $3^{-4} \times 3^{-2}$.

$$3^{-4} \times 3^{-2} = 3^{-4 + (-2)} = \mathbf{3^{-6}}$$

ANSWER: 3^{-6}

EXAMPLE 2 Simplify $5^4 \div 5^{-3}$.

$$5^4 \div 5^{-3} = 5^{4-(-3)} = \mathbf{5^7}$$

ANSWER: 5^7

Complete the following table.

	Meaning	In Words	Value
1. 7^{-2}	_____	_____	_____
2. 10^{-3}	_____	_____	_____
3. 3^{-4}	_____	_____	_____

Simplify each product or quotient.

4. $3^3 \times 3^{-1} =$ $4^{-2} \times 4^2 =$ $10^3 \times 10^{-4} =$

5. $6^2 \div 6^{-1} =$ $2^4 \div 2^{-2} =$ $5^3 \div 5^{-1} =$

Scientific Notation

Scientific notation is a shorthand way of writing numbers. In scientific notation, a number is written as the product of two factors. The first factor is a number between 1 and 10. The second factor is a power of 10. The power of 10 factor tells how many places to move the decimal point when changing from scientific notation to regular notation.

- A positive exponent means to move the decimal point to the right.
- A negative exponent means to move the decimal point to the left.

Positive Exponents	**Negative Exponents**

EXAMPLE 1 $7.5 \times 10^5 = 750,000$

move the decimal point five places to the right

EXAMPLE 2 $4.2 \times 10^{-3} = 0.0042$

move the decimal point three places to the left

In Scientific Notation	Number	In Scientific Notation	Number
3.8×10^3	3,800	5×10^{-2}	0.05
8.5×10^4	85,000	6×10^{-3}	0.006
6×10^7	60,000,000	7.5×10^{-5}	0.000075

Write each number in regular notation.

1. $6.3 \times 10^2 =$ $8.75 \times 10^6 =$ $9 \times 10^7 =$

2. $5.1 \times 10^{-2} =$ $8.5 \times 10^{-5} =$ $7 \times 10^{-6} =$

Using a positive exponent, write each number in scientific notation.

3. $5,000 =$ $30,000 =$ $75,000,000 =$

Using a negative exponent, write each number in scientific notation.

4. $0.03 =$ $0.0075 =$ $0.0000125 =$

Solve.

5. The approximate distance from Earth to the moon is 2.4×10^5 miles. Write this distance as a whole number.

6. The diameter of a human hair is measured to be 0.0012 inches. Write this diameter in scientific notation.

What Is a Square Root?

An important skill is finding a **square root** of a positive number. For example, to find the square root of 25, ask, "What number times itself equals 25?"

The number 25 has two square roots.

- 5 is a square root of 25. $5^2 = 25$

- -5 is a square root of 25. $(-5)^2 = 25$

The square root symbol is $\sqrt{}$.

Thus, $\sqrt{25} = +5$ and -5, or $\sqrt{25} = \pm 5$.

> A positive number has two square roots.
> - a positive square root (+)
> - a negative square root (−)
> - ± means "both positive and negative."
> EXAMPLE $\sqrt{36} = \pm 6$

EXAMPLE 1   7 is the square root of what number?

Since $7^2 = 49$, 7 is the square root of 49.

ANSWER: $7 = \sqrt{49}$

EXAMPLE 2   Find $\sqrt{64}$.

$8^2 = 64$ and $(-8)^2 = 64$.

ANSWER: $\sqrt{64} = \pm 8$

Numbers that have whole number square roots are called **perfect squares.** Perfect squares are easily found by squaring whole numbers. The first fifteen perfect squares are shown in the table below.

Table of Perfect Squares

$1^2 = 1$	$6^2 = 36$	$11^2 = 121$
$2^2 = 4$	$7^2 = 49$	$12^2 = 144$
$3^2 = 9$	$8^2 = 64$	$13^2 = 169$
$4^2 = 16$	$9^2 = 81$	$14^2 = 196$
$5^2 = 25$	$10^2 = 100$	$15^2 = 225$

As you already know, 0 times 0 is 0. Therefore, $0^2 = 0$.
To find the square root of a perfect square from the table,

- find the perfect square in the right-hand column

- read the whole number square root in the left-hand column

(**Remember:** Each perfect square has both a positive and a negative square root. For example, $\sqrt{121} = \pm 11$)

From the table on the previous page, find the square root of each perfect square below.

1. $\sqrt{169} = \pm\mathbf{13}$ $\sqrt{25} =$ $\sqrt{81} =$

2. $\sqrt{4} =$ $\sqrt{121} =$ $\sqrt{36} =$

3. $\sqrt{49} =$ $\sqrt{1} =$ $\sqrt{225} =$

4. $\sqrt{144} =$ $\sqrt{64} =$ $\sqrt{9} =$

5. $\sqrt{100} =$ $\sqrt{16} =$ $\sqrt{196} =$

··

The square root of a fraction is equal to the square root of the numerator over the square root of the denominator.

EXAMPLE 3   Find $\sqrt{\dfrac{36}{49}}$.

   STEP 1   Write as the square root of the numerator over the square root of the denominator.

   STEP 2   Find the square roots of both the numerator and the denominator.

$$\sqrt{\dfrac{36}{49}} = \dfrac{\sqrt{36}}{\sqrt{49}}$$

$$= \pm\dfrac{6}{7}$$

ANSWER: $\pm\dfrac{6}{7}$ **Check:** $\left(\dfrac{6}{7}\right)^2 = \dfrac{36}{49}$ and $\left(-\dfrac{6}{7}\right)^2 = \dfrac{36}{49}$

··

Find the square root of each fraction below. The first problem is done as an example.

6. $\sqrt{\dfrac{4}{9}} = \pm\dfrac{2}{3}$ $\sqrt{\dfrac{9}{64}} =$ $\sqrt{\dfrac{25}{36}} =$

7. $\sqrt{\dfrac{1}{4}} =$ $\sqrt{\dfrac{49}{81}} =$ $\sqrt{\dfrac{121}{144}} =$

8. $\sqrt{\dfrac{81}{16}} =$ $\sqrt{\dfrac{1}{196}} =$ $\sqrt{\dfrac{100}{121}} =$

Finding an Approximate Square Root

A number that is not a perfect square does not have a whole number square root. For example, the square root of 30 is between the whole numbers 5 and 6.

$$\sqrt{30} \text{ is greater than 5, since } 5^2 = 25$$

$$\sqrt{30} \text{ is less than 6, since } 6^2 = 36$$

To find an approximate square root of a number that is not a perfect square, follow these three steps.

STEP 1 Choose a perfect square that is close to the square root you want to find.

STEP 2 Divide this chosen number into the number you're trying to find the square root of.

STEP 3 Average your choice from Step 1 with the answer from Step 2. The average of these two numbers is an approximate square root.

(**Note:** The ≈ sign means "is approximately equal to.")

EXAMPLE 1 Find an approximate positive square root of 30.

STEP 1 Find the perfect square that is closest to the number 30. This perfect square is 25.

$$5 \approx \sqrt{30}$$
because $5 = \sqrt{25}$

STEP 2 Divide 5 into 30.

$$30 \div 5 = 6$$

STEP 3 Find the average of the numbers 5 and 6. The average is the approximate square root of 30.

$$5 + 6 = 11$$
$$11 \div 2 = \mathbf{5\tfrac{1}{2}}$$

ANSWER: $5\tfrac{1}{2}$ or 5.5   **Check:** $(5.5)^2 = 30.25$

EXAMPLE 2 Find an approximate positive square root of 54.

STEP 1 Find the perfect square that is closest to the number 54. Choose $7 \approx \sqrt{54}$ because $7 = \sqrt{49}$.

$$7 \approx \sqrt{54}$$

STEP 2 Divide 7 into 54.

$$54 \div 7 \approx 7.7$$

STEP 3 Find the average of the numbers 7 and 7.7.

$$7 + 7.7 = 14.7$$
$$14.7 \div 2 \approx \mathbf{7.4}$$

ANSWER: 7.4   **Check:** $(7.4)^2 = 54.76$

Find an approximate positive square root for each number. Round a decimal answer to the nearest tenth at each step.

1. $\sqrt{42} \approx$

2. $\sqrt{19} \approx$

3. $\sqrt{76} \approx$

4. $\sqrt{90} \approx$

5. $\sqrt{28} \approx$

6. $\sqrt{50} \approx$

7. $\sqrt{137} \approx$

8. $\sqrt{208} \approx$

Powers and Roots Review

Solve the problems below. When you finish, check your answers at the back of the book. Then correct any errors.

1. In the expression 3^4, _____ is the base, and _____ is the exponent.

Write each product as a base and exponent.

2. $(-4)(-4)(-4)(-4) =$

3. $(+5)(+5)(+5) =$

4. $\left(-\frac{3}{5}\right)\left(-\frac{3}{5}\right) =$

5. $8 \cdot 8 =$

6. $\left(\frac{2}{3}\right)\left(\frac{2}{3}\right)\left(\frac{2}{3}\right) =$

7. $9 \times 9 \times 9 \times 9 =$

Find each value.

8. $9^2 =$

9. $(-5)^3 =$

10. $(-3)^4 =$

11. $3^0 =$

12. $7^1 =$

13. $\left(\frac{2}{3}\right)^2 =$

14. $5^2 - (-3)^2 + 3^0 =$

15. $\left(\frac{3}{2}\right)^2 + \left(\frac{1}{4}\right)^1 - \left(\frac{1}{2}\right)^0 =$

16. $\frac{4^3}{3^2} =$

17. $\frac{3^2 \cdot 2^3}{4^2 \cdot 5^1} =$

Simplify each product or quotient.

18. $(5^4)(5^2)(5^3) =$

19. $\frac{8^6}{8^3} =$

Find each value.

20. $4^{-3} =$

21. $2^4 \cdot 2^{-3} =$

22. $10^{-1} \div 10^2 =$

Write each number in regular notation.

23. $4.5 \times 10^4 =$

24. $2.3 \times 10^{-3} =$

Write each number in scientific notation.

25. $720,000 =$

26. $0.006 =$

Find each square root.

27. $\sqrt{36} =$

28. $\sqrt{\dfrac{49}{64}} =$

29. Find an approximate square root of 56.

30. Find an approximate square root of 83.

ALGEBRAIC EXPRESSIONS AND FORMULAS

Writing Algebraic Expressions

In algebra, letters called **variables** or **unknowns** are used to represent numbers. A variable may be used in addition, subtraction, multiplication, or division problems. A variable may appear alone or it may be multiplied by a number called a **coefficient**.

$$3x + 4 \leftarrow \text{number standing alone}$$
↑↑——variable
coefficient

An **algebraic expression** consists of variables and numbers combined with one or more of the operations of addition, subtraction, multiplication, or division.

Operation	Algebraic Expression	Word Expression
Addition	$x + 4$	x plus 4
Subtraction	$y - 6$	y minus 6
	or	
	$6 - y$	6 minus y
Multiplication	$5z$	5 times z
Division	$\frac{n}{2}$	n divided by 2
	or	
	$\frac{2}{n}$	2 over n

Write an algebraic expression for each word expression.

1. n minus 8 _____

2. s divided by 5 _____

3. y plus 12 _____

4. w subtract 4 _____

5. 7 times x _____

6. r added to 19 _____

Write a word expression for each algebraic expression.

7. $5x$ _____

8. $4 + n$ _____

9. $7 - n$ _____

10. $\frac{m}{2}$ _____

11. $\frac{a}{5}$ _____

12. $s - 3$ _____

Using Parentheses with Algebraic Expressions

Many algebraic expressions contain more than one operation. The expression $2x + 8$ involves multiplication *and* addition. An algebraic expression may also contain parentheses. Placing a number outside of parentheses means that the whole expression is to be multiplied by the number.

Look at the difference between the expressions $3y + 7$ and $3(y + 7)$ when y is equal to 2.

Without parentheses: $3y + 7 = 3(2) + 7 = 6 + 7 = \mathbf{13}$
With parentheses: $3(y + 7) = 3(2 + 7) = 3(9) = \mathbf{27}$

Parentheses change the meaning and value of an expression. Below are more examples.

Algebraic Expression	Word Expression
$2y - 7$	2y minus 7
$2(y - 7)$	2 times the quantity y minus 7
$\frac{2}{3}x + 5$	$\frac{2}{3}$x plus 5
$\frac{2}{3}(x + 5)$	$\frac{2}{3}$ times the quantity x plus 5
$3x^2 + 1$	3 times x-squared plus 1
$3(x^2 + 1)$	3 times the quantity x-squared plus 1

Match each algebraic expression with its equivalent word expression.

Algebraic Expressions	Equivalent Word Expressions
_____ 13. $4x - 9$	**a.** z-squared minus 3
_____ 14. $-2a(a + 7)$	**b.** 4 times the quantity x minus 9
_____ 15. $z^2 - 3$	**c.** −2 times a-squared plus 7
_____ 16. $\frac{3}{4}x + 4$	**d.** $\frac{3}{4}$ times the quantity x plus 4
_____ 17. $z(z - 3)$	**e.** 4x minus 9
_____ 18. $4(x - 9)$	**f.** $\frac{3}{4}$x plus 4
_____ 19. $-2a^2 + 7$	**g.** z times the quantity z minus 3
_____ 20. $\frac{3}{4}(x + 4)$	**h.** −2a times the quantity a plus 7

Picturing an Algebraic Expression

An algebraic expression can be pictured on a line drawing.

EXAMPLE   The distance $\overline{AC}$ is the sum of the distances $\overline{AB}$ and $\overline{BC}$. This sum can be written as an algebraic expression.

$\overline{AC} = \overline{AB} + \overline{BC} = 2x + 6$

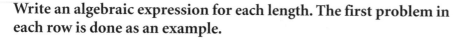

$\overline{AC} = 2x + 6$

Becoming familiar with algebraic expressions is an important algebra skill.

Write an algebraic expression for each length. The first problem in each row is done as an example.

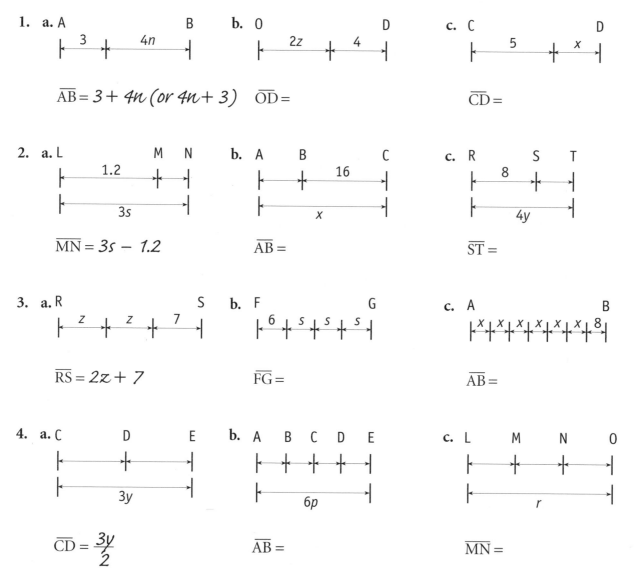

1. a. A   3   4n   B

    $\overline{AB} = 3 + 4n$ (or $4n + 3$)

    b. 0   2z   4   D

    $\overline{OD} =$

    c. C   5   x   D

    $\overline{CD} =$

2. a. L   1.2   M   N   3s

    $\overline{MN} = 3s - 1.2$

    b. A   B   16   C   x

    $\overline{AB} =$

    c. R   8   S   T   4y

    $\overline{ST} =$

3. a. R   z   z   7   S

    $\overline{RS} = 2z + 7$

    b. F   6   s   s   s   G

    $\overline{FG} =$

    c. A   x x x x x x   8   B

    $\overline{AB} =$

4. a. C   D   E   3y

    $\overline{CD} = \dfrac{3y}{2}$

    b. A   B   C   D   E   6p

    $\overline{AB} =$

    c. L   M   N   0   r

    $\overline{MN} =$

Simplifying an Algebraic Expression

Often, an algebraic expression can be simplified by combining **like terms.** Like terms have the same variable and exponent. For example, $5x$ and $3x$ are like terms; $3x^2$ and $2x^2$ are like terms. Numbers standing alone, such as 3 and 6 are also like terms.

To simplify an expression, add or subtract like terms.

Algebraic Expression	Simplified Expression
$5x + 3x$	$8x$
$7y - 4y + 5$	$3y + 5$
$2z + z + 3 + 6$	$3z + 9$
$3x^2 + 7x + 2x^2 - 3x$	$5x^2 + 4x$

Write a simplified algebraic expression for the perimeter of each triangle. (Remember: Perimeter is the distance around a figure.)

1. $P =$ _____

2. $P =$ _____

3. $P =$ _____

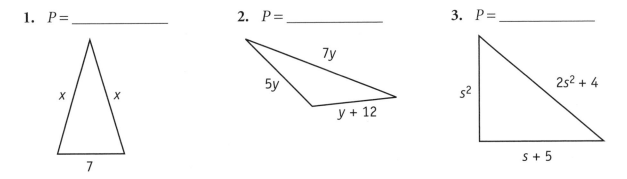

Write a simplified algebraic expression for the perimeter of each rectangle. (Remember: Perimeter of a rectangle is 2 lengths plus 2 widths.)

4. $P =$ _____

5. $P =$ _____

6. $P =$ _____

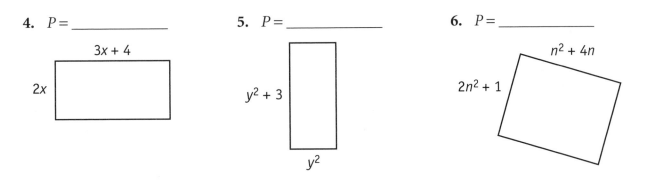

Algebraic Expressions and Word Problems

Algebraic expressions are often part of word problems.

EXAMPLE 1 Suppose *d* represents the amount that Joclyn earns each hour babysitting. Last Saturday Joclyn babysat for 6.5 hours. Write an algebraic expression that tells how much Joclyn earned last Saturday.

ANSWER: 6.5*d* (hours times *d*)

EXAMPLE 2 Mannie and two friends share the cost of a large pizza. Let *c* represent the price of the pizza. Write an algebraic expression that tells Mannie's share of the bill?

ANSWER: $\frac{c}{3}$ (cost divided by 3)

For problems 1–8, choose the correct algebraic expression.

1. On his last birthday, Arnie was 144 centimeters tall. Since that time, he has grown *h* centimeters. Which expression tells Arnie's present height?

   a. $h - 144$ **b.** $144 - h$ **c.** $h + 144$ **d.** $144h$

2. Shelley, Lauren, and Amy are going to share the cost of Sunday's breakfast. If the total bill is *t*, which expression tells each person's share?

   a. $\frac{t}{3}$ **b.** $t + 3$ **c.** $3t$ **d.** $t - 3$

3. Mai bought a magazine for $2.95. If she paid the clerk with *x* dollars, which expression tells how much change Mai would receive?

   a. $\$2.95 - x$ **b.** $\$2.95x$ **c.** $x + \$2.95$ **d.** $x - \$2.95$

4. Blair had a bag of *n* jelly beans. He gave 5 jelly beans to Andre and then divided the remaining candy among three other friends. Which expression tells how many jelly beans each of the other friends got?

   a. $\frac{(n+5)}{3}$ **b.** $\frac{(n-5)}{3}$ **c.** $3(n+5)$ **d.** $3(n-5)$

5. Christy lost 5% of her body weight while on a diet. If Christy weighed *n* pounds before dieting, which expression tells her weight now?

   a. $n + 0.05$ **b.** $n - 0.05$ **c.** $n + 0.05n$ **d.** $n - 0.05n$

6. If you subtract 6 from the square of Gary's age, the result is Zane's age. If Gary is g years old, which expression tells Zane's age?

    a. $g - 6^2$ b. $(g - 6)^2$ c. $g^2 - 6$ d. $(6 - g)^2$

7. If you multiply the square of a number n by 6 and then add 3 to the product, the sum is equal to 57. Which expression below equals 57?

    a. $6n^2 + 3$ b. $6(n^2 + 3)$ c. $6(n + 3)^2$ d. $(6n)^2 + 3$

8. Heidi has several coins. If you subtract 3 from the number she has and then square the difference, the result tells the number of dollar bills she has. If Heidi has n coins, which expression tells how many dollar bills she has?

    a. $n^2 - 3$ b. $n(n - 3)^2$ c. $(n - 3)^2$ d. $(n - 3)^2 - n$

For problems 9–14, write an algebraic expression as indicated.

9. Kelli is 154 centimeters tall. Suppose she grows h centimeters this next year. Write an expression that tells her height at the end of the year.

10. Marcel and four friends agree to share the cost renting several video games. The rental cost is n dollars. Write an expression that tells Marcel's share.

11. A sweater, regularly selling for p dollars, is on sale for 30% off. Write an expression that tells the savings being offered on the sweater.

12. Caren bought a jacket on sale. She paid 25% less than original price of m dollars. Write an expression that tells the amount that Caren paid.

13. If you multiply the square of a number n by 3 and then subtract 4 from the product, the difference equals Ben's age. Write an expression that tells Ben's age.

14. Wayne describes an algebraic expression. He says, "Subtract 12 times a number b from twice the square of the number." Write Wayne's expression.

Evaluating Algebraic Expressions

You find the value of an algebraic expression by substituting a number for each variable and then doing the operations in the following order.

1. Evaluate expressions within parentheses.
2. Evaluate powers.
3. Multiply and divide.
4. Add and subtract.

EXAMPLE 1 Find the value of $x + 2$ when $x = 3$.

    STEP 1 Substitute 3 for x. $\qquad\qquad\qquad$ $x + 2 = 3 + 2$
    STEP 2 Add. $\qquad\qquad\qquad\qquad\qquad\quad$ $= \mathbf{5}$

ANSWER: 5

EXAMPLE 2 Find the value of rt ($r \cdot t$) when $r = 40$ and $t = 3$.

    STEP 1 Substitute 40 for r and 3 for t. $\qquad$ $rt = 40 \cdot 3$
    STEP 2 Multiply. $\qquad\qquad\qquad\qquad\qquad$ $= \mathbf{120}$

ANSWER: 120

EXAMPLE 3 Find the value of $3y - 7$ when $y = 5$.
                (Multiply before subtracting.)

    STEP 1 Substitute 5 for y. $\qquad\qquad\qquad$ $3y - 7 = 3(5) - 7$
    STEP 2 Multiply. $\qquad\qquad\qquad\qquad\qquad$ $= 15 - 7$
    STEP 3 Subtract. $\qquad\qquad\qquad\qquad\qquad$ $= \mathbf{8}$

ANSWER: 8

EXAMPLE 4 Evaluate $4(x - y)$ when $x = 6$ and $y = 3$.

    STEP 1 Substitute 6 for x and 3 for y. $\quad$ $4(x - y) = 4(6 - 3)$
    STEP 2 Evaluate the expression within parentheses. $\quad$ $= 4(3)$
    STEP 3 Multiply. $\qquad\qquad\qquad\qquad\qquad$ $= \mathbf{12}$

ANSWER: 12

EXAMPLE 5 Evaluate $2(x + y)^2 - 4x$ when $x = 5$ and $y = -2$.

    STEP 1 Substitute 5 for x and -2 for y. $\quad$ $2(x + y)^2 - 4x = 2(5 + -2)^2 - 4(5)$
    STEP 2 Evaluate the parentheses. $\qquad\qquad\qquad\qquad$ $= 2(3)^2 - 4(5)$
    STEP 3 Evaluate the power. $\qquad\qquad\qquad\qquad\qquad$ $= 2(9) - 4(5)$
    STEP 4 Multiply. $\qquad\qquad\qquad\qquad\qquad\qquad$ $= 18 - 20$
    STEP 5 Subtract. $\qquad\qquad\qquad\qquad\qquad\qquad$ $= \mathbf{-2}$

ANSWER: −2

Find the value of each algebraic expression.

1. $x + 9$ when $x = -4$ $y - 12$ when $y = 3$ $z + (-5)$ when $z = 9$

2. $5n$ when $n = 4$ $7r$ when $r = -2$ $\frac{2}{3}s$ when $s = 24$

3. $\frac{x}{5}$ when $x = 35$ $\frac{y}{3}$ when $y = -18$ $\frac{a}{b}$ when $a = -10$,
        $b = -2$

4. xy when $x = -2$, mn when $m = -3$, uv when $u = 4$,
        $y = 4$ $n = -2$ $v = -2$

5. $12 + 5a$ when $a = 9$ $6x - 12$ when $x = 2$ $25 - xy$ when $x = 4$,
        $y = 3$

6. $ab - \frac{a}{b}$ when $a = 6$ $4x + \frac{x}{3}$ when $x = -9$ $\frac{c}{d} + 3c - 2d$ when $c = 8$,
        $b = 2$ $d = 4$

7. $3(x + 7)$ when $x = 2$ $-4(y - 5)$ when $y = 3$ $7(a - b)$ when $a = 2$,
        $b = 3$

8. $2x^2 + 3x$ when $x = 3$ $a^2 + b^2$ when $a = 6$, $3(c - d)^2 + 2c$ when $c = 9$,
        $b = 8$ $d = 8$

In an expression involving division, the division bar separates the problem into two parts. Evaluate the numerator and the denominator separately before dividing. Be sure to follow the order of operations shown on page 44.

EXAMPLE 6 Find the value of $\dfrac{3(y-4)^2}{2(y-1)}$ when $y = 6$.

STEP 1	Substitute 6 for y.	$\dfrac{3(y-4)^2}{2(y-1)} = \dfrac{3(6-4)^2}{2(6-1)}$
STEP 2	Evaluate both parentheses.	$= \dfrac{3(2)^2}{2(5)}$
STEP 3	Evaluate the power.	$= \dfrac{3(4)}{2(5)}$
STEP 4	Multiply.	$= \dfrac{12}{10}$
STEP 5	Simplify the improper fraction.	$= \dfrac{6}{5} = 1\dfrac{1}{5}$

ANSWER: $1\dfrac{1}{5}$

··

Find the value of each algebraic expression.

9. $\dfrac{2(x+3)^2}{5(x-1)}$ when $x = 2$ $\qquad\qquad$ $\dfrac{3(m-n)^2}{2(m+n)}$ when $m = -1, n = 3$

10. $\dfrac{4z(z-4)^2}{5z^2}$ when $z = -1$ $\qquad\qquad$ $\dfrac{2x(x-1)^2}{3y(y+1)^2}$ when $x = 2, y = 1$

Evaluating Formulas

A **formula** is a mathematical rule involving an algebraic expression. For example, the formula for the area of a rectangle is

Area = length times width or $A = lw$.

In a formula, the quantity that you are trying to find is usually written to the left of the equals sign. The algebraic expression is written to the right.

To find a value using a formula, substitute numbers for variables in the algebraic expression and do the indicated operations.

EXAMPLE   Use the formula $A = lw$ to find the area of a rectangle that has a length of 8 feet and a width of 5 feet.

Substitute 8 for l and 5 for w in the formula $A = lw$:
$A = 8 \times 5 = 40$

ANSWER: $A = 40$ square feet

(**Note:** In area problems, the answers are in square units. In volume problems, the answers are in cubic units.)

Find each value as indicated.

1. *Description:*   Perimeter of a rectangle
   Formula: $P = 2(l + w)$
   Variables: l = length
                   w = width

   Find P when $l = 7$ feet and $w = 4$ feet.

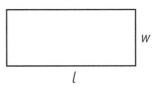

2. *Description:*   Perimeter of a triangle
   Formula: $P = a + b + c$
   Variables: a, b, c = sides of a triangle

   Find P when $a = 6$ feet, $b = 9$ feet, and $c = 13$ feet.

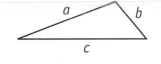

3. *Description:*   Circumference of a circle
   Formula: $C = 2\pi r$, where $\pi \approx \frac{22}{7}$
   Variables: r = radius

   Find C when $r = 7$ inches.

4. *Description:* Area of a rectangle
   Formula: $A = lw$
   Variables: l = length
          w = width

   Find A when $l = 12$ feet and $w = 7$ feet.

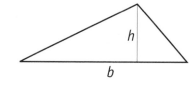

5. *Description:* Area of a triangle
   Formula: $A = \frac{1}{2}bh$
   Variables: b = base
          h = height

   Find A when $b = 14$ inches and $h = 6$ inches.

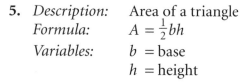

6. *Description:* Area of a circle
   Formula: $A = \pi r^2$, where $\pi \approx \frac{22}{7}$
   Variables: r = radius

   Find A when $r = 14$ inches.

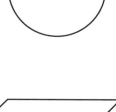

7. *Description:* Volume of a rectangular solid
   Formula: $V = lwh$
   Variables: l = length, w = width, h = height

   Find V when $l = 16$ inches, $w = 5$ inches, and $h = 9$ inches.

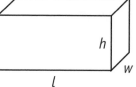

8. *Description:* Volume of a cylinder
   Formula: $V = \pi r^2 h$, where $\pi \approx \frac{22}{7}$
   Variables: r = radius, h = height

   Find V when $r = 2$ feet and $h = 7$ feet.

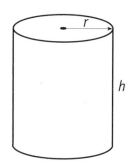

Formulas: Applying Your Skills

Using the formulas on the previous two pages, solve the problems below.

- Use the perimeter formulas for finding the distance around a figure.
- Use the circumference formula for finding the distance around a circle.
- Use area formulas for finding the surface area of a figure.
- Use volume formulas for finding the volume of (or the space inside) an object.

1. George is enclosing a rectangular window with molding. How many feet of molding are needed if the window measures 5 feet long by 4 feet wide?

2. A school playground in the shape of a triangle has sides of 150 yards, 125 yards, and 191 yards. What is the distance around the playground?

3. About how many feet of fence are needed to enclose a circular garden that has a radius of 28 feet?

4. Mary is putting new tile on her kitchen floor. How many square feet of tile does Mary need if the floor measures 15 feet long by 12 feet wide?

5. A triangular roof gable is 12 yards across and 3 yards high. Find the area of this gable.

6. A field in the shape of a circle has a radius of 21 meters. Find the approximate area of this field.

7. Carlos rented a moving van that has an enclosed section measuring 18 feet by 6 feet by 7 feet. How many cubic feet of storage space does the truck have?

Algebraic Expressions Review

Solve the problems below. When you finish, check your answers at the back of the book. Then correct any errors.

Write an algebraic expression for each word expression.

1. 7 subtracted from x _____

2. z divided by 9 _____

3. 19 added to y _____

4. -4 times n _____

5. 6 times the quantity x plus 4 _____

6. 2 times the square of the quantity $y - 5$ _____

Write a word expression for each algebraic expression.

7. $12 + y$ _____

8. $x - 9$ _____

9. $15a$ _____

10. $\dfrac{y}{12}$ _____

11. $4s + 8$ _____

12. $-9(2x + 4)$ _____

Write an algebraic expression as indicated.

13. A notebook is on sale for s dollars. Write an algebraic expression for the cost of six notebooks at this price.

14. Arlene paid m dollars for groceries that cost \$12.49. Write an algebraic expression that tells the amount of change Arlene should receive.

15. Janine took a ribbon that was n feet long and cut off a 2-foot piece. She then divided the remaining ribbon into six smaller pieces. Write an algebraic expression that tells the length of each smaller piece.

16. A pair of shoes is on sale for 40% off. The regular price of the shoes is n dollars. Write an algebraic expression that tells the sale price of the shoes.

Find the value of each algebraic expression.

17. $13 + 5x$ when $x = -3$

18. $7y - 8$ when $y = 6$

19. $\frac{z}{3} + 9$ when $z = -6$

20. $4xy$ when $x = 5$, $y = 3$

21. $-3(z + 7)$ when $z = -4$

22. $2a^2 - 4ab$ when $a = 2$, $b = 3$

23. $(m + n)(m - n)$ when $m = -4$, $n = 2$

24. $\dfrac{2(x + y)^2}{4(x - y)}$ when $x = 3$, $y = 1$

Solve.

25. Use the formula $V = \pi r^2 h$, where $\pi \approx \frac{22}{7}$, to estimate V when $r = 3$ feet and $h = 10$ feet.

26. In the formula $i = prt$, find i when $p = \$550$, $r = 0.09$, and $t = 1.5$.

27. Howard's kitchen is in the shape of a rectangle that is 15 feet long by 12 feet wide. Use the area formula $A = lw$ to find how many square feet of tile Howard needs to tile his kitchen.

28. Use the temperature formula $°C = \frac{5}{9}(°F - 32)$ to find the Celsius temperature (°C) when the Fahrenheit temperature (°F) is 104°F.

ONE-STEP EQUATIONS

What Is an Equation?

An **equation** is a statement that two quantities are equal. You may see an equation written in words or in mathematical symbols.

Equation in words: Two plus three is equal to five.
Equation in symbols: $2 + 3 = 5$

(**Note:** The equals sign = is read "is equal to" or simply "equals.")

Equations are common in mathematics. In daily life, mathematical questions often start as word problems and are solved as equations. We put words into symbols because symbols are easier to work with. Each time you add, subtract, multiply, or divide, you use an equation.

Familiar Equations
- Addition Equation: $32 + 19 = 51$
- Subtraction Equation: $76 - 30 = 46$
- Multiplication Equation: $43 \times 17 = 731$
- Division Equation: $\frac{80}{5} = 16$

In algebra, an equation may contain one or more letters in place of numbers. Each letter stands for a number whose value may not yet be known.

Algebraic Equations
- Addition Equation: $x + 23 = 47$
- Subtraction Equation: $y - 52 = 15$
- Multiplication Equation: $9a = 72$
- Division Equation: $\frac{b}{5} = 16$

Read an algebraic equation as follows.

$x + 23 = 47$ is read "x plus 23 is equal to 47"
$y - 52 = 15$ is read "y minus 52 equals 15"
$9a = 72$ is read "9 times a equals 72"
$\frac{b}{5} = 16$ is read "b divided by 5 is equal to 12"

> There is more than one correct way to read an algebraic equation.

Circle the equation described by the words.

1. The product of nine and a number is equal to fourteen.

   a. $n + 9 = 14$ **b.** $9n = 14$ **c.** $14n = 9$

2. The quotient of a number divided by six equals fifty.

   a. $6n = 50$ **b.** $\frac{6}{n} = 14$ **c.** $\frac{n}{6} = 50$

Write an addition equation for each drawing.

3.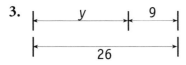

Equation: _____

4.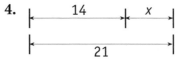

Equation: _____

Write a subtraction equation for each drawing.

5.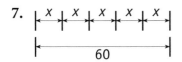

Equation: _____

6.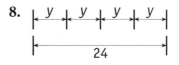

Equation: _____

Write a multiplication equation for each drawing.

7.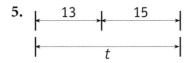

Equation: _____

8.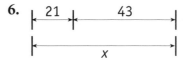

Equation: _____

Write a division equation for each drawing.

9.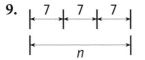

Equation: _____

10.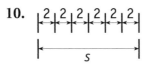

Equation: _____

Checking Unknowns

As you've seen, an algebraic equation contains a letter (variable) in place of a number. In an equation, this letter is often called an **unknown** because it is an unknown number.

Solving an equation means finding the value that makes the equation true. The **solution** is the value of the unknown that solves the equation.

To check if a value of the unknown is the correct solution, follow these two steps.

> **STEP 1** Substitute the value for the unknown into the equation.
>
> **STEP 2** Do the operations and compare each side of the equation.

<u>**EXAMPLE 1**</u>   Is $y = 5$ the solution for $3y - 9 = 6$?

> **STEP 1** Substitute 5 for y. $3(5) - 9 = 6$
>
> **STEP 2** Subtract. Compare. $15 - 9 = 6$
> $$6 = 6$$

Since $6 = 6$, $y = 5$ **is a solution** of the equation.

<u>**EXAMPLE 2**</u>   Is $x = 8$ the solution for $x + 7 = 14$?

> **STEP 1** Substitute 8 for x. $8 + 7 \neq 14$
>
> **STEP 2** Add. Compare. $15 \neq 14$

> The symbol $\neq$ means "is not equal to."

Since 15 is not equal to 14, $x = 8$ **is not a solution** of the equation.

Check to see if the suggested value is the solution to the equation. Circle Yes if the value is correct, No if it is not.

1. $x + 3 = 9$ Try $x = 4$.   Yes   No

2. $y - 12 = 37$ Try $y = 49$.   Yes   No

3. $3z = 39$ Try $z = 13$.   Yes   No

4. $\frac{a}{15} = 3$ Try $a = 54$.   Yes   No

5. $\frac{3}{4}m = 3$ Try $m = 4$.   Yes   No

6. $n + (-4) = 15$ Try $n = 11$.   Yes   No

7. $x - (-3) = 10$ Try $x = 7$.   Yes   No

8. $4s = -20$ Try $s = -5$.   Yes   No

Solving an Algebraic Equation

An algebraic equation is like a balance. The left side of the equation balances or equals the right side of the equation.

The addition equation $x + 3 = 5$ can be represented as weights on a balance. On the left side are the unknown weight x and 3 unit weights. On the right side are 5 unit weights.

To find x, *remove an equal number from each side* until x stands alone. Removing an equal number from each side ensures that the equation remains balanced.

You have solved the equation when x stands alone on the left and the value that balances x stands alone on the right.

The equation $x + 3 = 5$ is solved by *subtracting 3 from each side.* As shown at the right, $x = 2$.

Subtraction is used to solve the addition equation $x + 3 = 5$. Because subtraction *undoes* addition, subtraction is called the **inverse** (opposite) of addition.

As you'll see in the pages ahead, all algebraic equations are solved by using **inverse operations.** An inverse operation *undoes* an operation and leaves the unknown standing alone.

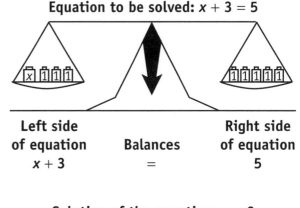

Equation to be solved: $x + 3 = 5$

Left side of equation	Balances	Right side of equation
$x + 3$	$=$	5

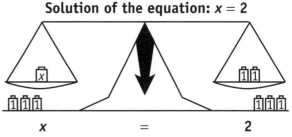

Solution of the equation: $x = 2$

| x | $=$ | 2 |

Operation		Inverse Operation	Solving an Equation
addition	$\leftrightarrow$	subtraction	Use subtraction to solve an addition equation.
subtraction	$\leftrightarrow$	addition	Use addition to solve a subtraction equation.
multiplication	$\leftrightarrow$	division	Use division to solve a multiplication equation.
division	$\leftrightarrow$	multiplication	Use multiplication to solve a division equation.

The example $x + 3 = 5$ also points out the most important rule to remember as you use inverse operations to solve equations:

Rule: Any inverse operation performed on one side of the equation must also be performed on the other side of the equation.

Solving an Addition Equation

To solve an addition equation, subtract the *added number* from each side of the equation. The unknown will be left alone, and the equation is solved.

EXAMPLE   Solve for x: $x + 6 = 15$ **Solve:** $x + 6 = 15$

   STEP 1   Subtract 6 from each side of the equation. $x + 6 - 6 = 15 - 6$

   STEP 2   Simplify both sides of the equation. $x = 9$
            On the left, 6 – 6 = 0, leaving x alone.
            On the right, 15 – 6 = 9.

ANSWER: $x = 9$

Check the answer by substituting 9 for x.   **Check:** $9 + 6 = 15$
                                                              ✓ $15 = 15$

Solve each addition equation and check each answer. The first problem in each row is partially completed. Show all steps.

1. $x + 5 = 8$ $x + 7 = 12$ $y + 3 = 9$ $n + 10 = 24$
$x + 5 - 5 = 8 - 5$

2. $y + \$2 = \$15$ $s + \$6 = \$11$ $x + \$5 = \$28$ $z + 17¢ = 35¢$
$y + \$2 - \$2 = \$15 - \$2$

3. $x + \frac{1}{2} = 4$ $y + \frac{2}{3} = 3$ $c + \frac{3}{4} = \frac{5}{4}$ $r + 1\frac{5}{8} = 5\frac{7}{8}$

$x + \frac{1}{2} - \frac{1}{2} = 4 - \frac{1}{2}$

4. $x + 3.2 = 6.8$ $n + 2.4 = 5.7$ $z + 5.5 = 14$ $y + 2.75 = 7.5$
$x + 3.2 - 3.2 = 6.8 - 3.2$

Solving a Subtraction Equation

To solve a subtraction equation, add the *subtracted number* to each side of the equation. The unknown will be left alone, and the equation is solved.

<u>EXAMPLE</u>   Solve for *x*: $x - 7 = 12$ **Solve:** $x - 7 = 12$

STEP 1   Add 7 to each side of the equation. $x - 7 + 7 = 12 + 7$

STEP 2   Simplify both sides of the equation. $x = 19$
On the left, $-7 + 7 = 0$, leaving *x* alone.
On the right, $12 + 7 = 19$.

ANSWER: *x* = 19

Check the answer by substituting 19 for *x*.   **Check:** $19 - 7 = 12$
$\checkmark 12 = 12$

Solve each subtraction equation and check each answer. The first problem in each row is partially completed. Show all steps.

1. $x - 6 = 14$ $x - 3 = 7$ $y - 6 = 10$ $z - 15 = 32$
$x - 6 + 6 = 14 + 6$

2. $y - \$5 = \$12$ $x - \$8 = \$18$ $y - \$11 = \$26$ $n - 9¢ = 27¢$
$y - \$5 + \$5 = \$12 + \$5$

3. $x - \frac{2}{3} = 6$ $r - \frac{5}{8} = 3$ $y - \frac{1}{4} = \frac{7}{4}$ $v - 2\frac{1}{2} = 5\frac{1}{2}$

$x - \frac{2}{3} + \frac{2}{3} = 6 + \frac{2}{3}$

4. $m - 2.4 = 5.3$ $x - 1.7 = 6.4$ $n - 7.5 = 9.75$ $y - 3.25 = 8.5$
$m - 2.4 + 2.4 = 5.3 + 2.4$

Word Problem Skills [+ and −]

On this page, you'll see how algebra equations can be written for problems you can easily solve without algebra. Practicing algebra skills here, though, is an excellent way to become more confident with the use of equations.

Circle the correct equation. Then find the value of the unknown.

1. If you subtract 18 from a number, the difference is 14. If n stands for the unknown number, which equation can be used to find n?

    a. $18 - n = 14$
    b. $n - 14 = 18$
    c. $n - 18 = 14$

    $n =$

2. Darrell spent $5.75 for lunch and had $2.15 left over. Which equation can be used to find the amount (m) Darrell had before lunch?

    a. $m - \$5.75 = \$2.15$
    b. $m - \$2.15 = \$5.75$
    c. $m + \$2.15 = \$5.75$

    $m =$

3. Lynn bought a pair of shoes that were on sale for 25% off the regular price of $40. Which equation can be used to find the amount Lynn paid (p)?

    a. $p + 0.25(\$40) = \$40$
    b. $p - 0.25(\$40) = \$40$
    c. $p - \$40 = 0.25(\$40)$

    $p =$

4. The temperature in Chicago rose 15°F between 9:00 A.M. and noon. If the temperature at noon is 87°F, which equation can be used to find the temperature (t) at 9:00 A.M.?

    a. $t - 15°F = 87°F$
    b. $t - 87°F = 15°F$
    c. $t + 15°F = 87°F$

    $t =$

Write an equation for each problem. Then solve your equation to find the unknown.

5. Ayla bought a sweater on sale for $34.00, $14.95 less than its original price. Write an equation to find the original price (p) of the sweater?

    Equation:

    $p =$

6. Blake had a full bag of cookies. After giving 14 cookies away, he still had 19 left. Write an equation to find the number (n) of cookies that were in the bag.

    Equation:

    $n =$

Solving a Multiplication Equation

To solve a multiplication equation, divide each side of the equation by the number that multiplies the unknown. The unknown will be left alone, and the equation is solved.

EXAMPLE   Solve for x: $4x = 32$

> **Solve:** $4x = 32$

STEP 1   Divide each side of the equation by 4.

$$\frac{\cancel{4}x}{\cancel{4}_1} = \frac{32}{4}$$

STEP 2   Simplify each side of the equation.

$$x = 8$$

On the left, $\frac{4}{4} = 1$, leaving x alone.

On the right, $\frac{32}{4} = 8$.

ANSWER: $x = 8$

Check the answer by substituting 8 for x.

> **Check:** $4(8) = 32$
> $\checkmark 32 = 32$

Solve each multiplication equation and check each answer. The first problem in each row is partially completed. Show all steps.

1. $7x = 35$ $8n = 48$ $6y = 54$ $7m = 49$

$$\frac{7x}{7} = \frac{35}{7}$$

2. $-4y = 24$ $-3x = 27$ $-5b = 65$ $-9y = 99$

$$\frac{-4y}{-4} = \frac{24}{-4}$$

3. $6x = \$18$ $8n = \$56$ $7z = \$147$ $11y = 121¢$

$$\frac{6x}{6} = \frac{\$18}{6}$$

4. $7y = 16.8$ $5y = 32.5$ $4.2x = 22.68$ $8.5n = 63.75$

$$\frac{7y}{7} = \frac{16.8}{7}$$

Solving a Division Equation

To solve a division equation, multiply each side of the equation by the number that divides the unknown. The unknown will be left alone, and the equation is solved.

__EXAMPLE__   Solve for x: $\frac{x}{6} = 14$

            STEP 1   Multiply each side of the equation by 6.

            STEP 2   Simplify each side of the equation.

                      On the left, $\frac{6}{6} = 1$, leaving x alone.

                      On the right, $14(6) = 84$.

Solve: $\frac{x}{6} = 14$

            $\frac{x}{6}(6) = 14(6)$

            $x = 84$

ANSWER: $x = 84$

    Check the answer by substituting 84 for x.

Check: $\frac{84}{6} = 14$

            $\checkmark 14 = 14$

Solve each division equation and check each answer. The first problem in each row is partially completed. Show all steps.

1. $\frac{x}{5} = 4$ $\frac{y}{3} = 9$ $\frac{c}{8} = 11$ $\frac{n}{4} = 12$

      $\frac{x}{5}(5) = 4(5)$

2. $\frac{y}{3} = \$6$ $\frac{x}{4} = \$17$ $\frac{n}{3} = \$12$ $\frac{t}{12} = \$1.50$

      $\frac{y}{3}(3) = \$6(3)$

3. $\frac{x}{2} = 3\frac{1}{2}$ $\frac{y}{3} = 2\frac{1}{3}$ $\frac{n}{6} = 2.5$ $\frac{r}{4} = 3.375$

      $\frac{x}{2}(2) = 3\frac{1}{2}(2)$

Word Problem Skills (× and ÷)

On this page, you'll practice using algebraic equations to solve multiplication and division problems.

Circle the correct equation. Then find the value of the unknown.

1. If you multiply a number n by 14, the product is 168. Which equation below can be used to find n?

   a. $168n = 14$

   b. $\frac{14}{n} = 168$

   c. $14n = 168$

   $n =$

2. On his diet, Brett lost 5% of his body weight. If Brett lost 9 pounds, which equation can be used to find his body weight (w) before he started dieting?

   a. $\frac{w}{0.05} = 9$

   b. $0.05w = 9$

   c. $9w = 0.05$

   $w =$

3. One out of four students in Marie's pottery class are men. If 7 students in the class are men, which equation can be used to find the number (n) of students in the class?

   a. $\frac{n}{4} = 7$

   b. $7n = \frac{1}{4}$

   c. $\frac{n}{7} = 4$

   $n =$

4. Kami and two friends are sharing the cost of lunch. If the total cost is $16.45, which equation can be used to find Kami's share (s)?

   a. $\frac{s}{3} = \$16.45$

   b. $s = \frac{\$16.45}{3}$

   c. $s = \frac{3}{\$16.45}$

   $s =$

Write an equation for each problem. Then solve your equation to find the unknown.

5. For every 5 hours Ayeisha spends in school, she does an hour of homework. This week she did 6.5 hours of homework. How many hours (h) did she spend in school this week?

   Equation:

   $h =$

6. One out of eight of the students in Ms. Knapp's kindergarten class are sick with the flu. If 4 students are sick, how many students (s) are in Ms. Knapp's class?

   Equation:

   $s =$

Solving an Equation with a Fraction Coefficient

To solve an equation with a **fraction coefficient,** multiply each side of the equation by the **reciprocal** of the fraction. The unknown will be left alone, and the equation is solved. The reciprocal is found by interchanging the numerator and the denominator.

EXAMPLE   Solve for x: $\frac{2}{5}x = 16$ **Solve:** $\frac{2}{5}x = 16$

STEP 1   Multiply each side of the equation by $\frac{5}{2}$. $\frac{5}{2}(\frac{2}{5})x = \frac{5}{2}(16)$

STEP 2   Simplify each side of the equation. $x = 40$

On the left, $\frac{5}{2}(\frac{2}{5}) = 1$, leaving x alone.

On the right, $\frac{5}{2}(16) = 40$.

ANSWER: $x = 40$

Check the answer by substituting 40 for x.   **Check:** $\frac{2}{5}(40) = 16$

$\checkmark 16 = 16$

Solve each equation and check each answer. The first problem in each row is partially completed. Show all steps.

1. $\frac{5}{8}x = 5$ $\frac{5}{6}z = 10$ $\frac{3}{7}a = 12$ $\frac{2}{3}n = 8$

$\frac{8}{5}(\frac{5}{8})x = \frac{8}{5}(5)$

2. $-\frac{3}{7}n = \frac{3}{4}$ $-\frac{1}{2}s = \frac{1}{2}$ $-\frac{3}{4}x = \frac{2}{3}$ $-\frac{2}{5}y = \frac{7}{8}$

$-\frac{7}{3}(-\frac{3}{7})n = -\frac{7}{3}(\frac{3}{4})$

3. $\frac{2}{3}z = \$3.50$ $\frac{3}{4}x = \$1.20$ $\frac{3}{8}c = \$6.60$ $\frac{5}{6}r = \$1.75$

$\frac{3}{2}(\frac{2}{3})z = \frac{3}{2}(\$3.50)$

Word Problem Skills (Fraction Coefficient)

On this page, you'll practice using algebraic equations that involve fraction coefficients.

Circle the correct equation. Then find the value of the unknown.

1. If you multiply a number x by $\frac{2}{3}$, the product is 36. Which equation below can be used to find x?

   a. $x \div \frac{2}{3} = 36$

   b. $\frac{2}{3}x = 36$

   c. $36x = \frac{2}{3}$

   $x =$

2. Chris saves $\frac{2}{5}$ of his monthly income. If he saves \$350 per month, which equation can be used to find his monthly income (i)?

   a. $\frac{2}{5}i = \$350$

   b. $\frac{3}{5}i = \$350$

   c. $i \div \frac{2}{5} = \$350$

   $i =$

3. One-fourth of the students in Lysha's class speak Spanish. If the class contains 18 students who do *not* speak Spanish, which equation can be used to find the total number of students (n) in the class?

   a. $\frac{1}{4}n = 18$

   b. $\frac{2}{4}n = 18$

   c. $\frac{3}{4}n = 18$

   $n =$

4. At a sale, Yvette paid \$7.75 for an umbrella labeled 20% off. Which equation can be used to find the regular price (r)?

   a. $\frac{1}{5}r = \$7.75$

   b. $\frac{4}{5}r = \$7.75$

   c. $\frac{6}{5}r = \$7.75$

   $r =$

Write an equation for each problem. Then solve your equation to find the unknown.

5. Three-fifths of Armand's work time is spent doing office work. If Armand does office work 21 hours each week, how many hours (h) does he work each week?

   Equation:

   $h =$

6. Five-eighths of the cars sold at Frank's Auto are foreign-made. If Frank sold 145 foreign-made cars last year, how many cars (c) did he sell in all?

   Equation:

   $c =$

One-Step Equations Review

Solve the problems below. When you finish, check your answers at the back of the book. Then correct any errors.

Solve each addition equation by subtraction.

1. $x + 8 = 10$

2. $z + (-5) = 8$

3. $a + 5 = 2$

Solve each subtraction equation by addition.

4. $y - 7 = 12$

5. $x - 12 = 13$

6. $b - 3 = -7$

Solve each multiplication equation by division.

7. $9z = 72$

8. $4x = -32$

9. $-13y = 143$

Solve each division equation by multiplication.

10. $\frac{x}{3} = 9$

11. $\frac{y}{7} = 8$

12. $\frac{a}{6} = -4$

Solve each equation with a fraction coefficient by multiplying by the reciprocal of the coefficient.

13. $-\frac{2}{5}x = 6$

14. $\frac{4}{9}y = 16$

15. $\frac{7}{4}a = -14$

Solve each equation.

16. $y - 12 = 5$

17. $\frac{4}{7}n = 20$

18. $5q = 120$

19. $\frac{y}{5} = 7$

20. $4n = -28$

21. $x + 10 = 7$

22. $6p = 42$

23. $x + (-7) = 18$

24. $s - 9 = -9$

25. $x - \frac{1}{6} = \frac{1}{2}$

26. $\frac{r}{3} = 2.25$

27. $5x = \frac{1}{3}$

28. $3k = 9.36$

29. $\frac{x}{3} = \frac{2}{5}$

30. $n - \frac{2}{3} = 1\frac{1}{3}$

31. $\frac{5}{6}z = 35$

32. $y - 3.25 = 8$

33. $x + \frac{3}{4} = 5\frac{1}{4}$

For each problem
a. **choose a letter to represent the unknown**
b. **write an equation using the unknown**
c. **solve the equation**

The first problem is done as an example.

34. Zeta spent $4.75 for breakfast and had $2.40 left over. How much money did Zeta have before paying for breakfast?

   a. Unknown: m = money Zeta had
   b. Equation: $m - \$4.75 = \$2.40$
   c. Solution:

$$m - \$4.75 + \$4.75 = \$2.40 + \$4.75$$
$$m = \$7.15$$

$$\overset{1}{\$4.75}$$
$$\underline{+\ \ 2.40}$$
$$\$7.15$$

35. Heather divided a box of cookies among five people. If each person received six cookies, how many cookies were in the box?

   a. Unknown:
   b. Equation:
   c. Solution:

36. Ephran received $2.11 in change when paying for a baseball. If the ball cost $7.89, how much money did Ephran give the clerk?

   a. Unknown:
   b. Equation:
   c. Solution:

37. Three-fourths of the students in Jenni's gymnastics class are girls. If there are 18 girls in the class, how many students are in the class?

   a. Unknown:
   b. Equation:
   c. Solution:

38. One out of six teachers at Grant School drive a foreign-made car. If four teachers drive a foreign-made car, how many teachers does Grant School have?

   a. Unknown:
   b. Equation:
   c. Solution:

39. Jocelyn saves $\frac{1}{3}$ of the money she earns babysitting and spends the rest. If she spent $90 during May, how much did Jocelyn earn babysitting during May?

   a. Unknown:
   b. Equation:
   c. Solution:

Write an equation for each drawing. Solve your equation. (Note: More than one equation is possible, but only one solution is correct.)

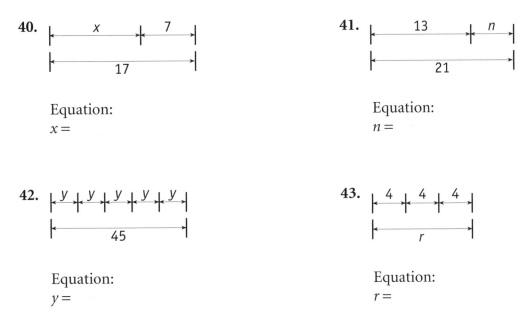

40.

Equation:

$x =$

41.

Equation:

$n =$

42.

Equation:

$y =$

43.

Equation:

$r =$

Write an equation for the perimeter of each triangle and solve for the missing side. (Remember: $P = s_1 + s_2 + s_3$). Problem 44 is started for you.

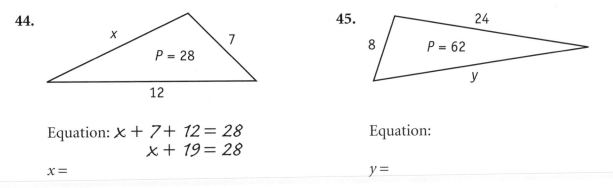

44.

Equation: $x + 7 + 12 = 28$
$x + 19 = 28$

$x =$

45.

Equation:

$y =$

Write an equation for the area of each rectangle and solve for the missing side. (Remember: $A = lw$) Problem 46 is started for you.

46.

Equation: $16w = 64$

$w =$

47.

Equation:

$l =$

MULTISTEP EQUATIONS

What Is a Multistep Equation?

A **multistep equation** is an equation that contains more than one operation. To solve a multistep equation, follow these two rules.

> **Rule 1:** Do addition or subtraction first.
> a. If a number is added to the unknown, subtract that number from each side of the equation.
> b. If a number is subtracted from an unknown, add that number to each side of the equation.

> **Rule 2:** Do multiplication or division last.
> a. If the unknown is multiplied by a number, divide each side of the equation by that number.
> b. If the unknown is divided by a number, multiply each side of the equation by that number.

EXAMPLE 1   Solve for x: $2x - 7 = 19$

  STEP 1   Add 7 to each side of the equation.

  STEP 2   Divide each side by 2.

Solve:
$$2x - 7 = 19$$
$$2x - 7 + 7 = 19 + 7$$
$$2x = 26$$
$$\tfrac{2}{2}x = \tfrac{26}{2}$$
$$x = 13$$

ANSWER: $x = 13$

  Check by substituting 13 for x.

Check: $2(13) - 7 = 19$
$$26 - 7 = 19$$
$$\checkmark 19 = 19$$

EXAMPLE 2   Solve for z: $\tfrac{z}{4} + 6 = 22$

  STEP 1   Subtract 6 from each side.

  STEP 2   Multiply each side by 4.

Solve:
$$\tfrac{z}{4} + 6 = 22$$
$$\tfrac{z}{4} + 6 - 6 = 22 - 6$$
$$\tfrac{z}{4} = 16$$
$$4\left(\tfrac{z}{4}\right) = 4(16)$$
$$z = 64$$

ANSWER: $z = 64$

  Check by substituting 64 for z.

Check:
$$\tfrac{64}{4} + 6 = 22$$
$$16 + 6 = 22$$
$$\checkmark 22 = 22$$

Solve each equation and check your answer. The first problem in each row is done as an example.

1. $5y + 4 = 19$ $\qquad$ $3x + 11 = 20$ $\qquad$ $2z + 9 = 1$

$5y + 4 - 4 = 19 - 4$

$\qquad 5y = 15$

$\qquad \dfrac{5y}{5} = \dfrac{15}{5}$

$\qquad\quad y = 3$

✓ $5(3) + 4 = 19$

$\quad\ 15 + 4 = 19$

$\qquad\ 19 = 19$

2. $3x - 9 = 3$ $\qquad$ $4z - 8 = 24$ $\qquad$ $-3a - 4 = 17$

$3x - 9 + 9 = 3 + 9$

$\qquad 3x = 12$

$\qquad \dfrac{3x}{3} = \dfrac{12}{3}$

$\qquad\ x = 4$

✓ $3(4) - 9 = 3$

$\quad\ 12 - 9 = 3$

$\qquad\ 3 = 3$

3. $\dfrac{y}{4} + 5 = 3$ $\qquad$ $\dfrac{z}{2} + 6 = 7$ $\qquad$ $-\dfrac{x}{5} + 3 = 9$

$\dfrac{y}{4} + 5 - 5 = 3 - 5$

$\qquad\quad \dfrac{y}{4} = -2$

$\qquad 4\left(\dfrac{y}{4}\right) = 4(-2)$

$\qquad\qquad y = -8$

✓ $-\dfrac{8}{4} + 5 = 3$

$\quad\ -2 + 5 = 3$

$\qquad\quad 3 = 3$

Learning a Shortcut to Solving Equations

To solve an equation, you must get the unknown standing alone. In the examples below, you'll learn a shortcut in which you do not write
- numbers that add to 0
- coefficients that equal 1

Shortcut   **Long Way**

EXAMPLE 1   Solve: $3x + 9 = 30$ Solve: $3x + 9 = 30$

STEP 1 Move 9 to the right side   **STEP 1** Subtract 9 from each side.
of the equation and change
the + sign to a − sign.

$3x = 30 - 9$ $3x + 9 - 9 = 30 - 9$
$3x = 21$ $3x = 21$

STEP 2 Divide 21 by 3.   **STEP 2** Divide each side by 3.

$x = \frac{21}{3}$ $\frac{3x}{3} = \frac{21}{3}$

ANSWER: $x = 7$ **ANSWER: $x = 7$**

EXAMPLE 2   Solve: $\frac{y}{2} - 12 = 31$ Solve: $\frac{y}{2} - 12 = 31$

STEP 1 Move 12 to the right side   **STEP 1** Add 12 to each side.
of the equation and change
the − sign to a + sign.

$\frac{y}{2} = 31 + 12$ $\frac{y}{2} - 12 + 12 = 31 + 12$

$\frac{y}{2} = 43$ $\frac{y}{2} = 43$

STEP 2 Multiply 43 by 2.   **STEP 2** Multiply each side by 2.

$y = 2(43) = \mathbf{86}$ $2\left(\frac{y}{2}\right) = 2(43)$

ANSWER: $y = 86$ **ANSWER: $y = 86$**

The steps of the shortcut can be summed up as follows.

STEP 1 If a number is added to (or subtracted from) an unknown, move the number to the other side of the equation and change the sign.

STEP 2 If a number multiplies an unknown, solve for the unknown by dividing the other side of the equation by the number.

STEP 3 If a number divides an unknown, solve for the unknown by multiplying the other side of the equation by the number.

Use the shortcut for the rest of the equations in this chapter.

Use the shortcut to solve each equation.

1. $3x + 11 = 29$

$$3x = 29 - 11$$
$$3x = 18$$
$$x = \frac{18}{3}$$
$$x = 6$$

$4y + 9 = 37$

$7z - 6 = 29$

2. $-12 + 2y = 14$

$$2y = 14 + 12$$
$$2y = 26$$
$$y = \frac{26}{2}$$
$$y = 13$$

$-5 + 7a = 30$

$13 - 9x = -41$

3. $\frac{y}{5} + 3 = 6$

$$\frac{y}{5} = 6 - 3$$
$$\frac{y}{5} = 3$$
$$y = 5(3)$$
$$y = 15$$

$\frac{z}{2} + 6 = 4$

$-\frac{n}{3} + 4 = 12$

4. $2x + 5 = 7$

$2 + 3a = 11$

$3z - 7 = 8$

5. $-14 + 2x = 4$

$\frac{z}{3} - 2 = 5$

$\frac{3}{4}n + 3 = 9$

Solving an Equation with Separated Unknowns

An equation is made up of **terms.** Each term is a number standing alone or an unknown multiplied by a coefficient.

When an unknown appears in more than one term, the separate terms can be combined. To combine terms, add or subtract the coefficients as indicated.

For example, the terms below are combined as follows.

$5x + 2x = 7x$	since $5 + 2 = 7$
$7y - 4y = 3y$	since $7 - 4 = 3$
$8z + z = 9z$	since $8 + 1 = 9$
$6n - n = 5n$	since $6 - 1 = 5$
$9w + (-5w) = 4w$	since $9 + (-5) = 4$

(**Note:** When no number multiplies (is in front of) an unknown, the coefficient is 1.)

To solve an equation in which the unknown appears in more than one term, combine the separate terms as your first step.

__EXAMPLE__   Solve for x: $4x - x + 2 = 17$

STEP 1   Combine the x's. $(4 - 1 = 3)$

STEP 2   Subtract 2 from 17.

STEP 3   Divide 15 by 3.

Solve:

$$4x - x + 2 = 17$$
$$3x + 2 = 17$$
$$3x = 17 - 2$$
$$3x = 15$$
$$x = \frac{15}{3}$$
$$x = 5$$

ANSWER: $x = 5$

Check by substituting 5 for x.

Check: $4(5) - 5 + 2 = 17$
$$20 - 5 + 2 = 17$$
$$\checkmark 17 = 17$$

Solve each equation.

1. $2x + 3x = 25$
    $$5x = 25$$
    $$x = \frac{25}{5}$$
    $$x = 5$$

$3y + 5y = 32$

$5z + z = 24$

2. $3a - a = 18 + 6$

$2a = 24$

$a = \dfrac{24}{2}$

$a = 12$

$7y - 3y = 23 + 9$

$13x - 8x = 17 + 8$

3. $2x + 4x - 9 = 15$

$6x - 9 = 15$

$6x = 15 + 9$

$6x = 24$

$x = \dfrac{24}{6}$

$x = 4$

$9z - z - 5 = 11$

$12y - 11y - 9 = 17$

4. $\dfrac{3}{4}y - \dfrac{1}{2}y - 5 = 7$

$\left(\dfrac{3}{4} - \dfrac{1}{2}\right)y - 5 = 7$

$\dfrac{1}{4}y = 7 + 5$

or $\dfrac{y}{4} = 12$

$y = 4(12)$

$y = 48$

$\dfrac{3}{5}a - \dfrac{1}{5}a - 3 = 17$

$\dfrac{2}{3}x + \dfrac{1}{6}x - 4 = 11$

5. $5z + 7z = 48$

$5y - 3y = 14$

$4a + 3a - 4 = 10$

6. $4a - 2a = 7 + 13$

$\dfrac{1}{2}x + \dfrac{1}{4}x = 9$

$2y - y + 2 = 7$

7. $5b - 2b = 14 - 5$

$3z + 6z + 7 = 29 - 4$

$y - \dfrac{1}{2}y = 3$

Terms on Both Sides of an Equation

When unknowns are on both sides of an equation, the first step is to move them to the same side, usually the left side. To move an unknown from one side of an equation to the other, simply move it and change the sign that precedes it.

__EXAMPLE 1__ Solve for x: $3x = 2x + 7$

    STEP 1 Move $2x$ to the left side of the equation, changing the sign from $+$ to $-$. (Subtract $2x$ from each side.)

    STEP 2 Subtract $2x$ from $3x$.

ANSWER: $x = 7$

Check by substituting 7 for x.

Solve: $\qquad 3x = 2x + 7$

$$3x - 2x = 7$$

$$x = 7$$

Check: $\qquad 3(7) = 2(7) + 7$

$$21 = 14 + 7$$

$$\checkmark\, 21 = 21$$

__EXAMPLE 2__ Solve for y: $2y - 6 = -3y + 24$

    STEP 1 Add $3y$ to the left side of the equation, changing the sign from $-$ to $+$.

    STEP 2 Combine the y's.

    STEP 3 Add 6 to 24.

    STEP 4 Divide 30 by 5.

ANSWER: $y = 6$

Check by substituting 6 for y.

Solve: $\qquad 2y - 6 = -3y + 24$

$$2y + 3y - 6 = 24$$

$$5y - 6 = 24$$

$$5y = 24 + 6$$
$$5y = 30$$

$$y = \frac{30}{5}$$
$$y = 6$$

Check: $\quad 2(6) - 6 = (-3)(6) + 24$
$$12 - 6 = -18 + 24$$
$$6 = 6$$

Solve each equation.

1. $4x = 3x + 6$ $5y = 2y + 9$ $7z = 5z + 18$

$$4x - 3x = 6$$
$$x = 6$$

2. $3z = 12 - z$

$3z + z = 12$

$4z = 12$

$z = \dfrac{12}{4}$

$z = 3$

$5a = 24 - 3a$

$7x = -27 - 2x$

3. $5x - 4 = 3x + 12$

$5x - 3x - 4 = 12$

$2x - 4 = 12$

$2x = 12 + 4$

$2x = 16$

$x = \dfrac{16}{2}$

$x = 8$

$9b - 7 = 5b + 25$

$8y - 9 = -y + 9$

4. $4m = 3m + 8$

$8x = 6x + 5$

$9y = 6y - 12$

5. $12x = 7x + 20$

$9y = 33 - 2y$

$3z = z - 18$

6. $4a + 17 = a - 13$

$6b - 12 = 2b - 4$

$z + 1 = \frac{1}{2}z + 4$

7. $5y + 9 = -2y + 30$

$14 + 2x = x + 9$

$11n - 3 = 9n + 3$

Multistep Equations: Word Problem Skills

Circle the correct equation. Then find the value of the unknown.

1. Eight times a number n plus 9 is equal to 73. Which equation can you use to find n?

    a. $8n - 9 = 73$
    b. $8n + 9 = 73$
    c. $9n + 8 = 73$

    $n =$

2. Five times a number w minus 7 is equal to three times the same number plus 19. Which equation is true for w?

    a. $5w + 7 = 3w + 19$
    b. $5w - 7 = 3w - 19$
    c. $5w - 7 = 3w + 19$

    $w =$

3. Susan and Terry run a day-care center at Susan's house. Susan earns twice as much as Terry. If they earn $225 during May, which equation can be used to find Terry's share (t)?

    a. $t + 2t = \$225$
    b. $t - 2t = \$225$
    c. $3t - t = \$225$

    $t =$

4. Lucy earns $400 a month in salary. She also receives an extra $18 for each appliance she sells. If last month Lucy earned a total of $886, how many appliances (a) did she sell?

    a. $\$400 - \$18a = \$886$
    b. $\$18a + \$886 = \$400$
    c. $\$18a + \$400 = \$886$

    $a =$

Write an equation for each problem. Then solve your equation to find the unknown.

5. A number n multiplied by 3 is equal to the same number added to 12. What is the number n?

    Equation:

    $n =$

6. Two-thirds of a number m plus one-sixth of the same number is equal to 25. What is the number m?

    Equation:

    $m =$

7. Denise pays a 6% sales tax in her state. Recently, Denise bought a new bed for a total price of $360.40, including tax. What was the price (p) of the bed before tax?

Equation:

$p =$

8. Joe has money in a savings account. If he adds $50 a month for 6 months, he will have three times the amount he has now, not counting the interest. How much savings (s) does Joe have in his account now?

Equation:

$s =$

Write an equation for each drawing. Solve your equation. (Note: More than one equation is possible, but only one solution is correct.)

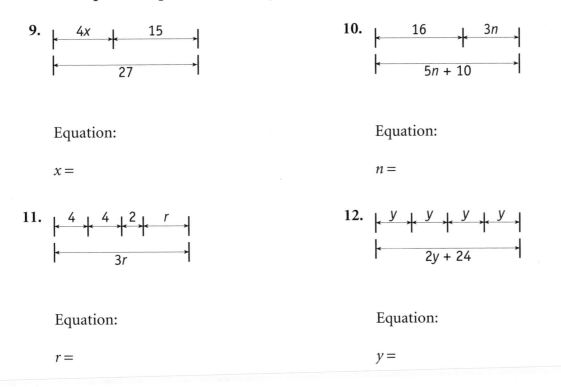

9.

4x   15

27

Equation:

$x =$

10.

16   3n

5n + 10

Equation:

$n =$

11.

4   4   2   r

3r

Equation:

$r =$

12.

y   y   y   y

2y + 24

Equation:

$y =$

Find the value of x in each triangle below. (Remember: $P = s_1 + s_2 + s_3$)

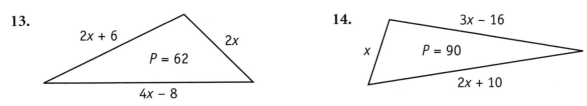

13.

2x + 6   2x

P = 62

4x − 8

$x =$

14.

3x − 16

x   P = 90

2x + 10

$x =$

Solving an Equation with Parentheses

Parentheses are commonly used in algebraic equations. Parentheses are used to identify terms (numbers or variables) that are to be multiplied by another term, usually a number.

The first step in solving an equation is to remove the parentheses by multiplication. Then, combine separated unknowns and solve for the unknown.

Follow these four steps to solve an algebraic equation.

STEP 1   Remove parentheses by multiplication.

STEP 2   Combine separated unknowns.

STEP 3   Do addition or subtraction first.

STEP 4   Do multiplication or division last.

To remove parentheses, multiply each term inside the parentheses by the term outside the parentheses. If the parentheses are preceded by a negative sign or number, remove the parentheses by changing the sign of each term within the parentheses.

<u>**EXAMPLES**</u> $4(x + 3) = 4x + 12$
$5(y - 4) = 5y - 20$
$-3(n + 3) = -3n - 9$
$-2(z - 1) = -2z + 2$

<u>**EXAMPLE 1**</u>   Solve for x: $4(x + 3) = 20$ **Solve:** $4(x + 3) = 20$

     STEP 1   Remove parentheses by $4x + 12 = 20$
                  multiplying each term.

     STEP 2   Subtract 12 from 20. $4x = 20 - 12$
                                                    $4x = 8$

     STEP 3   Divide 8 by 4. $x = \frac{8}{4}$

                                                        $x = 2$

ANSWER: $x = 2$

<u>**EXAMPLE 2**</u>   Solve for z: $4z - (3z + 2) = 5$ **Solve:** $4z - (3z + 2) = 5$

     STEP 1   Remove parentheses by $4z - 3z - 2 = 5$
                  changing the sign of each term
                  within parentheses.

     STEP 2   Combine the z's. $z - 2 = 5$

     STEP 3   Add 2 to 5. $z = 5 + 2$
                                                      $z = 7$

ANSWER: $z = 7$

Solve each equation. The first problem in each row is done as an example. (Remember: Remove the parentheses as your first step.)

1. $2(x+4)=20$

$2x+8=20$

$2x=20-8$

$2x=12$

$x=\dfrac{12}{2}$

$x=6$

$3(n+2)=27$

$5(b+3)=40$

2. $3(y-6)=15$

$3y-18=15$

$3y=15+18$

$y=\dfrac{33}{3}$

$y=11$

$2(a-3)=16$

$4(b-2)=8$

3. $-2(x-4)=-12$

$-2x+8=-12$

$-2x=-12-8$

$-2x=-20$

$x=\dfrac{-20}{-2}$

$x=10$

$-3(m-2)=-15$

$-2(2y+1)=6$

4. $4(y-3)=3(y+6)$

$4y-12=3y+18$

$4y-3y-12=18$

$y-12=18$

$y=18+12$

$y=30$

$5(z-1)=4(z+4)$

$7(a-1)=6(a+1)$

Equations with Parentheses: Word Problem Skills

In the following examples, notice how parentheses are used to represent information contained in word problems. Study these examples before solving the word problems on the next page.

EXAMPLE 1 Three times the quantity of a number minus 4 is equal to two times the sum of the number plus 3. What is the number?

> **STEP 1** Let $x =$ the unknown number
> $3(x - 4)$ is three times the quantity of x minus 4
> $2(x + 3)$ is two times the sum of x plus 3
>
> **STEP 2** Write an equation for the problem. **Solve:** $3(x - 4) = 2(x + 3)$
>
> **STEP 3** Solve the equation.
> **a.** Remove parentheses. $3x - 12 = 2x + 6$
> **b.** Move $2x$ to the left side $3x - 2x - 12 = 6$
> of the equation, changing
> the sign from + to –.
> **c.** Combine the x's. $x - 12 = 6$
> **d.** Move 12 to the right side of the equation. $x = 6 + 12$
> **e.** Add 12 to 6. $x = 18$

ANSWER: $x = 18$

EXAMPLE 2 Mary, Anne, and Sally share living expenses. Anne pays $25 less rent than Mary. Sally pays twice as much rent as Anne. If the total rent is $365, how much rent does each pay?

> **STEP 1** Let $x =$ Mary's rent (**Hint:** Since you don't know how much
> $x - 25 =$ Anne's rent Mary pays for rent, let her rent equal x.)
> $2(x - 25) =$ Sally's rent
>
> **STEP 2** Write an equation. Mary's + Anne's + Sally's = total rent
> **Solve:** $x + (x - 25) + 2(x - 25) = \$365$
>
> **STEP 3** Solve the equation.
>
> **a.** Remove parentheses. $x + x - 25 + 2x - 50 = \$365$
> **b.** Combine the x's $4x - 75 = \$365$
> and the numbers.
> **c.** Move 75 to the right side $4x = 365 + 75$
> of the equation.
> **d.** Add 75 to 365. $4x = 440$
> **e.** Divide 440 by 4. $x = \frac{440}{4}$
> $x = 110$

ANSWER: $x = \$110$, **Mary's rent**
$x - 25 = \$85$, **Anne's rent**
$2(x - 25) = \$170$, **Sally's rent**

Circle the correct equation. Then find the value of the unknown.

1. Three times the sum of a number n plus 1 equals two times the sum of the number plus 4. Which equation can be used to find n?

   a. $3(n + 1) = 2n + 4$
   b. $3(n + 1) = 2(n + 4)$
   c. $3(n + 4) = 2(n + 1)$

   $n =$

2. Maria, Amy, and Sadie share food costs. Amy pays \$10 a month less than Maria. Sadie pays twice as much as Amy. If the monthly food bill is \$310, which equation can be used to find Maria's share (m)?

   a. $m - (m + \$10) - 2(m + \$10) = \$310$
   b. $m + (m - \$10) - 2(m + \$10) = \$310$
   c. $m + (m - \$10) + 2(m - \$10) = \$310$

   $m =$

Write an equation for each problem. Then solve your equation to find the unknown.

3. Four times the sum of a number n plus 2 equals 3 times the sum of the number plus 5. What is n?

   Equation:

   $n =$

4. Two-thirds times the quantity y minus 3 is equal to one-third y. What is y?

   Equation:

   $y =$

5. Sami, Ben, and Louis went to lunch. Ben's meal cost \$0.45 less than Sami's. Louis's meal cost twice as much as Ben's. If the bill came to \$9.25, how much does each owe? (**Hint:** Let s = Sami's share)

   Equation:

   Sami owes:
   Ben owes:
   Louis owes:

6. Consecutive integers are whole numbers that follow one another. For example, 11, 12, and 13 are consecutive integers. If the sum of three consecutive integers is 54, what are the integers? (**Hint:** Let x = 1st integer)

   Equation:

   1st integer:
   2nd integer:
   3rd integer:

Multistep Equations Review

Solve the problems below. When you finish, check your answers at the back of the book. Then correct any errors.

Basic Equations

1. $3x + 14 = 62$

2. $4n + 9 = 25$

3. $r - 7 = 65$

4. $4y - 13 = 31$

5. $\frac{x}{2} + 5 = 17$

6. $\frac{2}{3}n - 4 = 12$

Equations with Separated Unknown

7. $4n + 3n = 35$

8. $5x - 2x + 9 = 36$

9. $\frac{2}{3}y - \frac{1}{6}y + 4 = 19$

Equations with Terms on Both Sides

10. $7x = 3x + 20$

11. $4n + 7 = 3n + 30$

12. $11y - 6 = 9y + 28$

Equations with Parentheses

13. $6(b - 2) = 12$

14. $4(z + 2) = 2z + 40$

15. $3(y + 2) = 2(y + 5)$

Mixed Practice

16. $5w - 2w = 39$

17. $3x + 5 = 20$

18. $4n = 2n + 14$

19. $5(x - 3) = 10$

20. $5s - 2s + 9 = 24$

21. $6r + 2 = 3r + 11$

22. $2(x + 5) = x + 4$

23. $4y - 8 = 32$

24. $\frac{n}{2} = n + 1$

25. $4(y + 3) = 2y$

26. $3(z - 2) = 2(z - 1)$

27. $\frac{3}{4}x - \frac{1}{4}x + 5 = 27$

28. $4p = 2p + 8$

29. $5n - 6 = 2n + 12$

30. $x - 2 = \frac{1}{2}x + 5$

Circle the correct equation. Then find the value of the unknown.

31. Six times a number *m* subtract 4 is equal to 26. Which equation can you use to find *m?*

    a. $6m - 26 = 4$
    b. $6m - 4 = 26$
    c. $4m - 6 = 26$

    $m =$

32. If Jolene adds five stamps each month for 8 months to her collection, she will have twice the number of stamps she has now. Which equation can you use to find the number of stamps (*n*) Jolene has now?

    a. $n - 40 = 2n$
    b. $2n + 8(5) = n$
    c. $n + 8(5) = 2n$

    $n =$

Write an equation for each problem. Then solve your equation to find the unknown.

33. Five times the quantity *x* minus 1 equals 3 times the quantity *x* plus 9. What is *x?*

    Equation:

    $x =$

34. Four times a number *n* minus 3 times the sum of the number plus 2 is equal to 5. What is the number?

    Equation:

    $n =$

35. Lateisha has a coin collection. If she adds five coins each month, the size of her collection will triple in 2 years. How many coins (*c*) does Lateisha have in her collection now?

    Equation:

    Number of Coins:

36. Alyce, Jon, and Kim worked together on a painting job. Jon worked 8 fewer hours than Alyce. Kim worked twice as many hours as Jon. If the total hours worked on the job was 72, how many hours did each person work? (**Hint:** Let *h* = hours worked by Alyce)

    Equation:

    Alyce:
    Jon:
    Kim:

37. Consecutive even integers are even whole numbers that follow one another. For example, 12, 14, and 16 are consecutive even integers. If the sum of three consecutive even integers is 66, what are the integers? (**Hint:** Let x = 1st number)

Equation:

1st number:

2nd number:

3rd number:

38. Find the value of x in the triangle below.

Equation:

$x =$

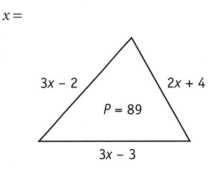

Write an equation for each drawing. Solve your equation. (Note: More than one equation is possible, but only one solution is correct.)

39.
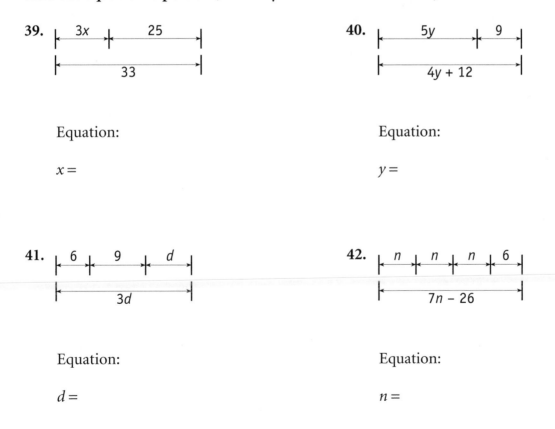

Equation:

$x =$

40.

Equation:

$y =$

41.

Equation:

$d =$

42.

Equation:

$n =$

SPECIAL EQUATIONS

Understanding Ratios

A **ratio** is a comparison of two numbers. For example, if there are 5 women and 4 men in a class, the *ratio of women to men* is 5 to 4.

You can write the ratio *5 to 4* in two ways.

- As a fraction, *5 to 4* is $\frac{5}{4}$.

- Using a colon, *5 to 4* is 5:4.

> Read the ratios $\frac{5}{4}$ and 5:4 as "five to four."

Write a ratio in the same order as it appears in a question. In the example above, the ratio of men to women is 4 to 5.

> To simplify a ratio, write it as a fraction and follow these three rules.
>
> **1.** Reduce a ratio to lowest terms.
> 5 to 10 = $\frac{5}{10}$ = $\frac{1}{2}$
>
> **2.** Leave an improper-fraction ratio as an improper fraction.
> 8 to 6 = $\frac{8}{6}$ = $\frac{4}{3}$
>
> **3.** Write a whole-number ratio as an improper fraction.
> 4 to 1 = $\frac{4}{1}$

Solve each ratio problem.

1. There are 12 girls and 16 boys in Kira's class. What is the ratio of boys to girls?

2. Chad's grandfather is 80 years old. Chad is 25 years old. What is the ratio of Chad's age to his grandfather's age?

3. Tiffani's soccer team won 12 games and lost 8.

   a. What is the ratio of games won to games lost?

   b. What is the ratio of games won to all games played?

4. Of the 18 teachers in Shane's school, 14 are women. What is the ratio of men teachers to women teachers at this school?

5. What is the ratio of the distance between Rose City and Chad to the distance between Rose City and Garrett?

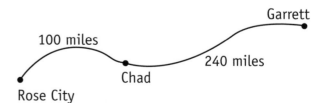

Using Equations to Solve Ratio Problems

Some ratio problems are most easily solved by using equations.

EXAMPLE   The ratio of two numbers is 4 to 3. The sum of the numbers
is 42. What are the two numbers?

STEP 1   Represent the greater number as $4x$ and the
lesser number as $3x$.

STEP 2   Set the sum of the numbers equal to 42.

$4x + 3x = 42$

STEP 3   Solve the equation for x.

$$4x + 3x = 42$$
$$7x = 42$$
$$x = \frac{42}{7}$$
$$x = 6$$

STEP 4   Determine the two numbers.

greater number $= 4x = 4(6) =$ **24**

lesser number $= 3x = 3(6) =$ **18**

> The key to solving this problem is choosing to represent the unknown numbers as $4x$ and $3x$— whole number terms that are in the ratio of 4 to 3.

ANSWER: 18 and 24

**Write an equation for each problem. Then solve your equation to find
the unknown.**

1. The ratio of two numbers is 3 to 2. The
    sum of the numbers is 75. What are the
    two numbers?

    Equation:

    first number:
    second number:

2. The ratio of two numbers is 5 to 2. The
    difference of the numbers is 18. What are
    the two numbers?

    Equation:

    first number:
    second number:

3. For every $4 Alan earns, Shannon earns
    $6. On a job that pays $740, how much
    will each person earn?

    Equation:

    Alan:
    Shannon:

4. A garden is in the shape of a rectangle. The
    ratio of the garden's length to its width is
    7 to 4. If the perimeter of the garden is
    176 feet, what is its length and width?

    Equation:

    length:
    width:

Understanding Proportions

A **proportion** is an equation made up of two equal ratios. For example, if you add 2 cups of orange juice to 3 cups of pineapple juice, the ratio of orange juice to pineapple juice is $\frac{2}{3}$. You make the same drink by adding 4 cups of orange juice to 6 cups of pineapple juice. The drinks are the same because the ratios of orange juice to pineapple juice are equal: $\frac{2}{3} = \frac{4}{6}$.

You read a proportion as two equal ratios connected by the word *as*.

$\frac{2}{3} = \frac{4}{6}$ is read "2 is to 3 *as* 4 is to 6."

> A proportion is usually written as a pair of equivalent (equal) fractions.

In a proportion, the cross products are equal. To find the cross products, multiply each numerator by the opposite denominator.

Cross Multiplication

$$\frac{2}{3} \diagup\!\!\!\!\diagdown \frac{4}{6}$$

Equal Cross Products

$$2(6) = 3(4)$$
$$12 = 12$$

Finding a Missing Number in a Proportion

To complete a proportion, write the letter n (or some other letter) for the missing number.

- Write the cross products.

- Solve the equation for the chosen variable.

EXAMPLE 1 Find the missing number: $\frac{n}{3} = \frac{8}{12}$

STEP 1 Cross multiply. $12n = 24 \ (3 \times 8)$

STEP 2 Divide each side by 12. $n = \frac{24}{12} = \mathbf{2}$

ANSWER: $n = 2$

EXAMPLE 2 Find the missing number: $\frac{4}{x} = \frac{20}{15}$

STEP 1 Cross multiply. $20x = 60 \ (4 \times 15)$

STEP 2 Divide each side by 20. $x = \frac{60}{20} = \mathbf{3}$

ANSWER: $x = 3$

Cross multiply to see if each pair of fractions forms a proportion.
Circle Yes if the cross products are equal, No if they are not.

1. $\frac{2}{3} \overset{?}{=} \frac{9}{12}$ Yes   No

   $\frac{3}{4} \overset{?}{=} \frac{6}{8}$ Yes   No

   $\frac{4}{5} \overset{?}{=} \frac{5}{4}$ Yes   No

2. $\frac{3}{7} \overset{?}{=} \frac{9}{21}$ Yes   No

   $\frac{9}{16} \overset{?}{=} \frac{5}{8}$ Yes   No

   $\frac{4}{5} \overset{?}{=} \frac{8}{10}$ Yes   No

Write cross products for each proportion. The first problem is done as an example.

3. $\frac{n}{2} = \frac{8}{16}$ $\frac{4}{12} = \frac{3}{n}$ $\frac{8}{20} = \frac{x}{5}$ $\frac{x}{3} = \frac{10}{12}$

   $16n = 2 \times 8$
   $16n = 16$

4. $\frac{10}{x} = \frac{5}{9}$ $\frac{n}{4} = \frac{24}{32}$ $\frac{8}{3} = \frac{x}{6}$ $\frac{n}{8} = \frac{18}{12}$

In each proportion below
- **write the cross products**
- **divide to find the value of n or x**

5. $\frac{n}{2} = \frac{7}{14}$ $\frac{x}{3} = \frac{12}{18}$ $\frac{6}{8} = \frac{x}{4}$ $\frac{5}{n} = \frac{10}{4}$

   $n =$ $x =$ $x =$ $n =$

6. $\frac{n}{6} = \frac{27}{18}$ $\frac{3}{8} = \frac{x}{32}$ $\frac{n}{10} = \frac{4}{5}$ $\frac{4}{x} = \frac{12}{9}$

   $n =$ $x =$ $n =$ $x =$

Using Proportions to Solve Word Problems

Proportions are used to solve word problems involving comparisons or rates. Remember to write the terms of a proportion in the order stated in the problem. As you'll see, numbers in a proportion are not always whole numbers.

EXAMPLE 1 A recipe calls for 2 cups of sugar for each 5 cups of flour. How many cups of sugar are needed for 14 cups of flour?

STEP 1 Write a proportion where each ratio is $\frac{\text{cups of sugar}}{\text{cups of flour}}$.

Let n stand for the unknown cups of sugar.

$$\frac{2}{5} = \frac{n}{14}$$

STEP 2 Write the cross products.

$$5n = 28 \ (14 \times 2)$$

STEP 3 Divide by 5 to solve for n.

$$n = \frac{28}{5} = 5\frac{3}{5}$$

ANSWER: $n = 5\frac{3}{5}$ cups of sugar

EXAMPLE 2 Working 7 hours, Blake earned $64. At this same pay rate, how much would Blake earn working 10 hours?

STEP 1 Write a proportion where each ratio is $\frac{\text{hours worked}}{\text{amount earned}}$.

Let n stand for the unknown earnings.

$$\frac{7}{\$64} = \frac{10}{n}$$

STEP 2 Write the cross products.

$$7n = \$640 \ (\$64 \times 10)$$

STEP 3 Divide by 7 to solve for n.

$$n = \frac{\$640}{7} = \$91.43$$

(to the nearest cent)

ANSWER: $n = \$91.43$

Circle the proportion that can be used to solve each problem.

1. On a map of Ohio, an actual distance of 75 miles is only 2 inches of map distance. On this map, how many inches (n) apart are two towns whose actual distance apart is 280 miles?

   a. $\frac{n}{2} = \frac{75}{280}$

   b. $\frac{75}{2} = \frac{280}{n}$

   c. $\frac{n}{75} = \frac{2}{280}$

2. When he painted his house, Bob mixed 2 pints of blue paint with each 3 gallons of white. How many pints (p) of blue paint did Bob use if he used a total of 19 gallons of white paint?

   a. $\frac{p}{3} = \frac{2}{19}$

   b. $\frac{p}{2} = \frac{3}{19}$

   c. $\frac{2}{3} = \frac{p}{19}$

3. Bonita drove 450 miles in 8 hours. At this same rate, how many hours (t) would it take Bonita to drive another 225 miles?

   a. $\dfrac{450}{8} = \dfrac{225}{t}$

   b. $\dfrac{t}{8} = \dfrac{450}{225}$

   c. $\dfrac{t}{450} = \dfrac{8}{225}$

4. In a survey of 280 high school students, 3 out of 5 said they plan to work during summer vacation. How many students (n) do *not* plan to work?

   a. $\dfrac{3}{5} = \dfrac{n}{280}$

   b. $\dfrac{n}{280} = \dfrac{3}{8}$

   c. $\dfrac{2}{5} = \dfrac{n}{280}$

Write a proportion as your first step in solving each problem. (Note: More than one correct proportion is possible, but only one answer is correct.)

5. If 8 ounces of steak costs $1.74, how much does 12 ounces of steak cost?

6. Jeni drove her car 102 miles on 3 gallons of gas. How far can Jeni expect to drive on 16 gallons of gas?

7. A light tan paint is made by mixing 2 parts of brown paint to 5 parts of white. How many quarts of brown must be mixed with 15 quarts of white to make the light tan color?

8. If 12.7 centimeters exactly equals 5 inches, how many centimeters exactly equals 2 inches?

9. Georgia earned $55.40 for 8 hours of work on Monday. At this rate, how much can Georgia earn if she works an additional 20 hours over the next few days?

10. Tough Stuff epoxy is to be mixed in a ratio of 3 parts hardener to 7 parts base. How many drops of hardener are needed for 28 drops of base.

Solving a Linear Equation

A **linear equation** is an equation that contains two variables where each variable has an exponent of 1. (**Remember:** An exponent of 1 is usually not written.) For example, $y = 2x + 3$ is a linear equation that contains the variables x and y.

A linear equation has more than one solution. For each value of x, there is a matching value of y. In $y = 2x - 3$, if you substitute 5 for x, y is equal to 7 [$2(5) - 3$].

To solve a linear equation follow these two steps.

STEP 1 Choose several values for the second variable.

STEP 2 Substitute each value for the second variable into the equation and find the matching value of the first variable.

To keep the solutions in neat order, you write them in a **Table of Values.**

<u>EXAMPLE 1</u> Solve for y: $y = 3x$, when $x = -2, 0, 2$, and 4

STEP 1 Write the chosen values for x in the Table of Values.

STEP 2 Substitute each value for x into the equation $y = 3x$.

STEP 3 Fill in the Table of Values for y.

x value	y = 3x	Table of Values x	y
$x = -2$	$y = 3(-2) = -6$	-2	-6
$x = 0$	$y = 3(0) = 0$	0	0
$x = 2$	$y = 3(2) = 6$	2	6
$x = 4$	$y = 3(4) = 12$	4	12

<u>EXAMPLE 2</u> Solve for y: $y = 2x - 3$, when $x = -1, 0$, and 3

STEP 1 Write the chosen values of x in the Table of Values.

STEP 2 Substitute each value of x into the equation $y = 2x - 3$.

STEP 3 Fill in the Table of Values for y.

x value	y = 2x - 3	Table of Values x	y
$x = -1$	$y = 2(\overset{-2}{\overbrace{(-1)}}) - 3 = -5$	-1	-5
$x = 0$	$y = 2(\overset{0}{\overbrace{(0)}}) - 3 = -3$	0	-3
$x = 3$	$y = 2(\overset{6}{\overbrace{(3)}}) - 3 = 3$	3	3

For each linear equation, complete the Table of Values.

1. $y = x + 4$ **Table of Values**

x	y
0	
1	
2	
3	

2. $y = 3x - 2$ **Table of Values**

x	y
1	
2	
3	
4	

3. $a = -2b + 3$ **Table of Values**

b	a
−1	
0	
1	
2	

4. $r = \frac{1}{2}s + 3$ **Table of Values**

s	r
−2	
0	
2	
4	

Complete the Table of Values as indicated.

5. The perimeter of the rectangle below is given by the equation $P = 2w + 6$. Complete the Table of Values for P.

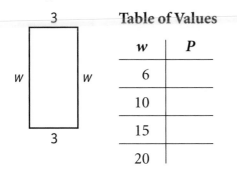

Table of Values

w	P
6	
10	
15	
20	

6. M and N are complementary angles. Use the equation $\angle M = 90° - \angle N$ to complete the Table of Values for $\angle M$.

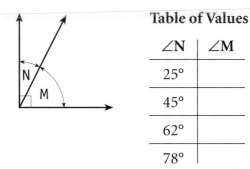

Table of Values

∠N	∠M
25°	
45°	
62°	
78°	

Solving a System of Equations

Two linear equations often have a **common solution**—an x and a y value that is a solution for each equation at the same time. Finding a common solution for two or more linear equations is called **solving a system of equations.** One way to find a common solution is to subtract one equation from the other.

EXAMPLE 1 What x value and y value is a common solution to the following equations?

$$\text{Equation 1: } y = 2x + 3$$
$$\text{Equation 2: } y = x + 2$$

STEP 1 Subtract Equation 2 from Equation 1. Subtract the y's on the left side. Subtract the x's and numbers on the right side.

$$\begin{aligned} y = 2x + 3 \quad &= \quad y = 2x + 3 \\ -(y = x + 2) \quad &= \quad \underline{- y = -x - 2} \\ &\qquad\quad 0 = x + 1 \end{aligned}$$

STEP 2 Solve for x by subtracting 1 from each side of the equation.

$$0 - 1 = x + 1 - 1$$
$$-1 = x$$
$$\text{or } x = -1$$

STEP 3 Substitute the x value (–1) into the second equation. Solve the equation to find the y value.

$$y = x + 2$$
$$= -1 + 2$$
$$y = 1$$

ANSWER: $x = -1$, $y = 1$

Check:

Equation 1	Equation 2
$y = 2x + 3$	$y = x + 2$
$1 = 2(-1) + 3$	$1 = -1 + 2$
$1 = -2 + 3$	$\checkmark 1 = 1$
$\checkmark 1 = 1$	

EXAMPLE 2 What x value and y value is a common solution to the following equations?

$$\text{Equation 1: } y = 2x + 1$$
$$\text{Equation 2: } y = -x + 4$$

STEP 1 Subtract Equation 2 from Equation 1.

$$\begin{aligned} y = 2x + 1 \quad &= \quad y = 2x + 1 \\ -(y = -x + 4) \quad &= \quad \underline{- y = -(-x) - 4} \\ &\qquad\quad 0 = 3x - 3 \end{aligned}$$

STEP 2 Solve for x by adding 3 to each side of the equation. Then divide each side by 3.

$$0 + 3 = 3x - 3 + 3$$
$$3 = 3x \text{ (or } 3x = 3)$$
$$\frac{3x}{3} = \frac{3}{3} \text{ (Divide by 3.)}$$
$$x = 1$$

STEP 3 Substitute the x value (1) into the second equation. Solve the equation to find the y value.

$$y = -x + 4$$
$$= -1 + 4$$
$$y = 3$$

ANSWER: $x = 1$, $y = 3$

Check:

Equation 1	Equation 2
$y = 2x + 1$	$y = -x + 4$
$3 = 2(1) + 1$	$3 = -1 + 4$
$3 = 2 + 1$	$\checkmark 3 = 3$
$\checkmark 3 = 3$	

Use subtraction to solve each system of equations.

1. $y = 2x + 3$
   $y = x + 1$

   $x =$
   $y =$

2. $y = 3x + 6$
   $y = 2x + 1$

   $x =$
   $y =$

3. $y = 5x - 3$
   $y = 2x$

   $x =$
   $y =$

4. $y = 4x - 1$
   $y = 3x + 1$

   $x =$
   $y =$

5. $y = 2x + 9$
   $y = -x$

   $x =$
   $y =$

6. $y = x - 3$
   $y = -x + 5$

   $x =$
   $y =$

Quadratic Equations

In a linear equation, such as $y = 2x + 2$, there are many possible solutions—in fact, an infinite number.

- For each value of x there is a single value of y.
- For each value of y there is a single value of x.

A **quadratic equation,** unlike a linear equation, contains the square of a variable. For example, $y = x^2$ is a quadratic equation. For $y = x^2$, there are also an infinite number of solutions. When solving a quadratic equation, you sometimes want to find x values when the y value is given. As a first step, the variable that is squared is written to the left of the equals sign. For example, the equation $y = x^2$ would be written as $x^2 = y$.

Notice that for each possible value of y, there are *two correct* values of x. One value is a positive square root; the other is a negative square root.

EXAMPLE 1 Solve the equation $x^2 = y$ when $y = 16$.

    STEP 1 Substitute 16 for the variable y.

$$x^2 = 16$$

    STEP 2 Solve for x by taking the square root of each side of the equation.

$$\sqrt{x^2} = \sqrt{16}$$
$$x = +4 \text{ or } -4$$

ANSWER: $x = \pm 4$

EXAMPLE 2 Solve the equation $3x^2 = y$ when $y = 12$.

    STEP 1 Substitute 12 for the variable y.

$$3x^2 = 12$$

    STEP 2 Divide each side of the equation by 3.

$$\frac{3x^2}{3} = \frac{12}{3}$$
$$x^2 = 4$$

    STEP 3 Solve for x by taking the square root of each side.

$$\sqrt{x^2} = \sqrt{4}$$
$$x = +2 \text{ or } -2$$

ANSWER: $x = \pm 2$

EXAMPLE 3 Solve the equation $2x^2 - 50 = y$ when $y = 0$.

    STEP 1 Substitute 0 for the variable y.

$$2x^2 - 50 = 0$$

    STEP 2 Add 50 to each side of the equation.

$$2x^2 - 50 + 50 = 0 + 50$$
$$2x^2 = 50$$

    STEP 3 Divide each side of the equation by 2.

$$\frac{2x^2}{2} = \frac{50}{2}$$
$$x^2 = 25$$

    STEP 4 Solve for x by taking the square root of each side of the equation.

$$\sqrt{x^2} = \sqrt{25}$$
$$x = +5 \text{ or } -5$$

ANSWER: $x = \pm 5$

Solving for *y*

For each value of *x*, find the single correct value of *y*. The first problem in each row is done as an example.

1. $y = x^2$ when $x = 3$
   $y = 3^2$
   $\quad = 9$

   $y = 2x^2$ when $x = -2$

   $y = \frac{3}{4}x^2$ when $x = 4$

2. $y = x^2 + 3$ when $x = 2$
   $y = 2^2 + 3$
   $\quad = 4 + 3$
   $\quad = 7$

   $y = x^2 - 7$ when $x = -3$

   $y = 2x^2 + 5$ when $x = 4$

Solving for *x*

For each value of *y*, find the *two* correct values of *x*. The first problem in each row is done as an example.

3. $x^2 = y$ when $y = 25$
   $x^2 = 25$
   $x = \pm 5$

   $3x^2 = y$ when $y = 48$

   $2x^2 = y$ when $y = 72$

4. $x^2 - 49 = y$ when $y = 0$
   $x^2 - 49 = 0$
   $\quad x^2 = 49$
   $\quad\quad x = \pm 7$

   $x^2 - 64 = y$ when $y = 0$

   $4x^2 - 144 = y$ when $y = 0$

5. $x^2 + 4 = y$ when $y = 13$
   $x^2 + 4 = 13$
   $\quad x^2 = 9$
   $\quad\quad x = \pm 3$

   $2x^2 = y$ when $y = 50$

   $x^2 - 50 = y$ when $y = 119$

Special Equations Review

Solve the problems below. When you finish, check your answers at the back of the book. Then correct any errors.

1. The Jets baseball team won 25 games and lost 10.

    a. What is the ratio of games won to games lost?

    b. What is the ratio of games won to number of games played?

Write an equation for each problem. Then solve your equation to find the unknown.

2. The ratio of two numbers is 3 to 4. The sum of the numbers is 63. What are the two numbers?

3. A soccer practice field is 3 times as long as it is wide. If the perimeter of the field is 176 yards, what is the length of the field?

Solve each proportion.

4. $\dfrac{n}{8} = \dfrac{9}{12}$

5. $\dfrac{3}{y} = \dfrac{18}{12}$

6. $\dfrac{25}{40} = \dfrac{20}{x}$

Write a proportion as your first step in solving each problem.

7. Regan earned $46.20 for 7 hours of work. At this rate, how much can Regan earn working 12 hours?

8. Three out of every 5 teachers at Jefferson High School are men. How many of Jefferson's 40 teachers are men?

For each linear equation, complete the Table of Values.

9. $y = x - 2$ **Table of Values**

x	y
0	
2	
4	
6	

10. $y = 2x + 3$ **Table of Values**

x	y
−1	
0	
1	
2	

Use subtraction to solve each system of equations.

11. $y = 2x + 5$
$y = x + 3$

$x =$
$y =$

12. $y = 3x + 1$
$y = x + 1$

$x =$
$y =$

For each value of *x*, find the single correct value of *y*.

13. $y = x^2$ when $x = 4$

14. $y = 2x^2$ when $x = -3$

For each value of *y*, find the *two* correct values of *x*.

15. $x^2 = y$ when $y = 36$

16. $4x^2 = y$ when $y = 64$

GRAPHING EQUATIONS

Becoming Familiar with a Coordinate Grid

A coordinate grid is formed by combining two number lines.
- The vertical number line is called the **y-axis.**
- The horizontal number line is called the **x-axis.**
- The point at which the two lines meet is called the **origin.**

Every point on a coordinate grid has coordinates, two numbers that tell its position.
- The **x-coordinate** tells how far the point is from the y-axis. Positive x indicates the point is to the right of the y-axis. Negative x indicates the point is to the left of the y-axis.
- The **y-coordinate** tells how far the point is from the x-axis. Positive y indicates the point is above the x-axis. Negative y indicates the point is below the x-axis.
- Coordinates are usually written as an **ordered pair,** two numbers within parentheses. The x-coordinate (x value) is written first, followed by the y-coordinate (y value).

__EXAMPLE__   On the grid below, point A is (−6, 4).

x-coordinate ⤒ ⤒ y-coordinate

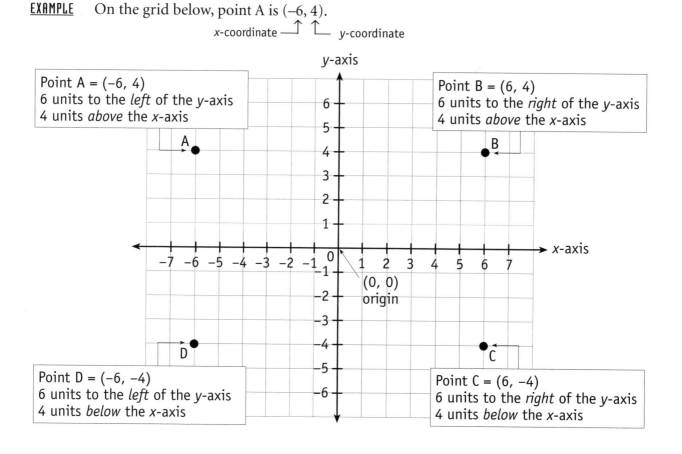

Point A = (−6, 4)
6 units to the *left* of the y-axis
4 units *above* the x-axis

Point B = (6, 4)
6 units to the *right* of the y-axis
4 units *above* the x-axis

Point D = (−6, −4)
6 units to the *left* of the y-axis
4 units *below* the x-axis

Point C = (6, −4)
6 units to the *right* of the y-axis
4 units *below* the x-axis

Write each pair of coordinates as an ordered pair. The first problem is
done as an example.

1. a. $x = 4$
        $y = -2$
        (4, –2)

    b. $x = -3$
        $y = 5$

    c. $x = 2$
        $y = 6$

    d. $x = -4$
        $y = -7$

Identify the *x*- and *y*-coordinates of each point. The first problem is
done as an example.

2. a. $R = (2, -8)$
        $x = $ **2**
        $y = $ **–8**

    b. $S = (-1, 6)$
        $x = $
        $y = $

    c. $T = (0, 5)$
        $x = $
        $y = $

    d. $U = (-4, 0)$
        $x = $
        $y = $

Write the coordinates of each point as an ordered pair. The first point
is done as an example.

3.

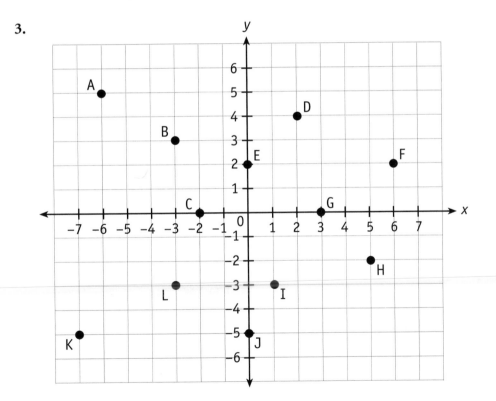

A = **(–6, 5)**   B =   C =   D =

E =   F =   G =   H =

I =   J =   K =   L =

Plotting Points on a Grid

To plot a point on a coordinate grid, follow these steps.

STEP 1 Locate the *x* value on the *x*-axis.

STEP 2 From the *x* value, move directly up (for a positive *y* value) or directly down (for a negative *y* value).

STEP 3 Label this point.

EXAMPLE 1   Plot point A = (5, −4) on the grid at the right.

STEP 1 Find the value *x* = 5 on the *x*–axis.

STEP 2 From the value 5 on the *x*-axis, move down to the *y* value −4.

STEP 3 Label point A.

EXAMPLE 2   Plot point B = (0, 5)

STEP 1 Begin at *x* = 0, the *x* value. Move up to the *y* value 5.

STEP 2 Label point B.

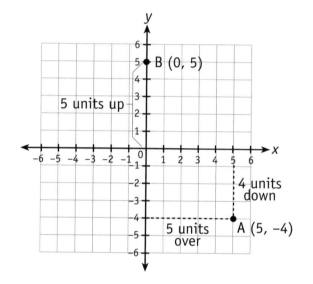

Plot and label each point.

1.

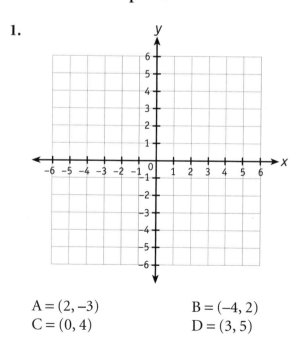

A = (2, −3)   B = (−4, 2)
C = (0, 4)   D = (3, 5)

2.

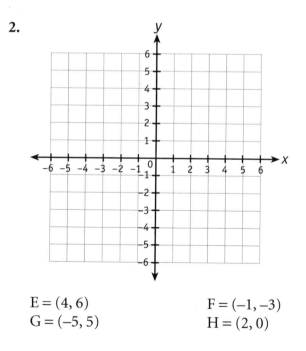

E = (4, 6)   F = (−1, −3)
G = (−5, 5)   H = (2, 0)

Graphing a Line

When plotted points lie on a straight line, draw a line through the points and extend this line to the edges of the grid. The points where the line crosses the axes are called **intercepts** of the line.

- The **x-intercept** is the point where the line crosses the x-axis. For the x-intercept, the y-coordinate is 0.
- The **y-intercept** is the point where the line crosses the y-axis. For the y-intercept, the x-coordinate is 0.

<u>EXAMPLE</u>   Graph the line containing the points $(-4, -2)$, $(0, 2)$, and $(2, 4)$.

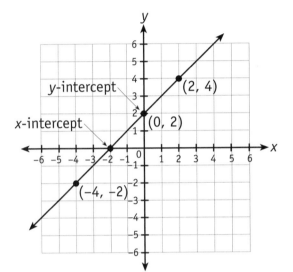

STEP 1   Plot the points and label each point with its coordinates.

STEP 2   Draw a line through the points. Extend the line to the edges of the grid.

STEP 3   Identify the points where the line crosses each axis.

x-intercept = (−2, 0)

y-intercept = (0, 2)

Graph the line containing each set of three points. Find the x- and y-intercepts of each line.

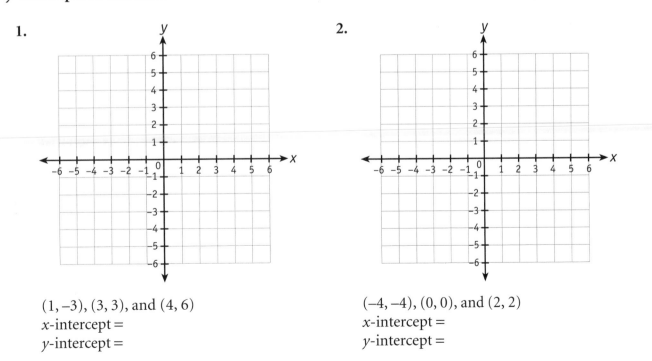

1.

$(1, -3)$, $(3, 3)$, and $(4, 6)$
x-intercept =
y-intercept =

2.

$(-4, -4)$, $(0, 0)$, and $(2, 2)$
x-intercept =
y-intercept =

Finding the Slope of a Line

You are already familiar with the concept **slope.** When you walk uphill, you walk up a slope. When you walk downhill, you walk down a slope. A graphed line, like a hill, has a slope.

The slope of a line is given as a number. When you move between two points on the line, the slope is found by dividing the change in *y* value by the corresponding change in *x* value.

> Slope of a line $= \dfrac{\text{Change in } y \text{ value (vertical change)}}{\text{Change in } x \text{ value (horizontal change)}}$

- A line that goes up from left to right has a *positive slope.*
- A line that goes down from left to right has a *negative slope.*

In Example 1, the slope of the line is 2.   In Example 2, the slope of the line is $-\left(\frac{2}{2}\right) = -1$.

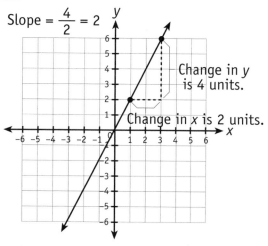

The line above has *positive slope* because it goes *up* from left to right.

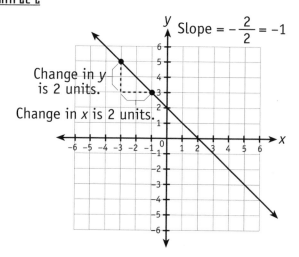

The line above has *negative slope* because it goes *down* from left to right.

Zero Slope and Undefined Slope

Two types of lines have neither positive nor negative slope.

- A horizontal line has *zero slope.* The *x*-axis (or any horizontal line) is a line with 0 slope.
- A vertical line has an *undefined slope.* The concept of slope does not apply to a vertical line. The *y*-axis (or any vertical line) is a line with undefined slope.

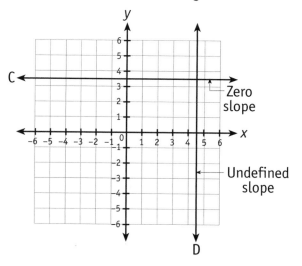

Solve the following problems.

1. Name the slope of each line as *positive, negative, zero,* or *undefined.*

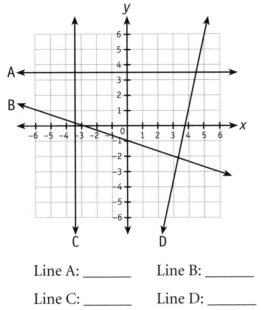

Line A: _____   Line B: _____

Line C: _____   Line D: _____

2. Find the numerical value of the slope of line E. Identify the *x*- and *y*-intercepts.

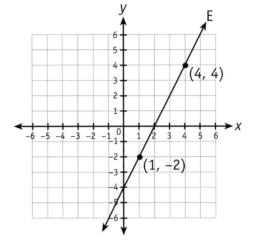

Slope of line E is _____.
　　　　　　　　　　　(number)

x-intercept = (　,　)

y-intercept = (　,　)

3. Find the numerical value of the slope of line F. Identify the *x*- and *y*-intercepts.

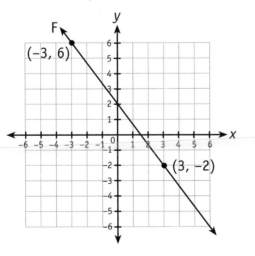

Slope of line F is _____.
　　　　　　　　　　　(number)

x-intercept = (　,　)

y-intercept = (　,　)

4. Subtract the coordinates to find the slope of the line that passes through each pair of points. Part a is done as an example.

a. (3, 1) and (4, 6)

$$\text{slope} = \frac{\text{change in } y}{\text{change in } x} = \frac{6-1}{4-3} = \frac{5}{1} = 5$$

b. (0, 2) and (1, 4)

c. (1, 2) and (−2, 5)

d. (0, 0) and (3, 2)

Graphing a Linear Equation

When solutions of a linear equation such as $y = 3x - 1$ are plotted on a coordinate grid, they always lie on a straight line. Drawing this line of solutions is called **graphing the equation.**

To graph a linear equation, follow these steps.

STEP 1 Choose three values for x and find the matching values for y. Write each pair of values in a Table of Values.

STEP 2 Plot the three points on a coordinate grid; then connect them with a line extending to the edges of the grid.

> **Note:** Only two points are needed to draw a line, but the third point serves as a check that you plotted the first two points correctly.

EXAMPLE Graph the equation $y = 3x - 1$.

STEP 1 Let $x = 0$, 1, and 2. Solve the equation $y = 3x - 1$ for these three values. Make a Table of Values.

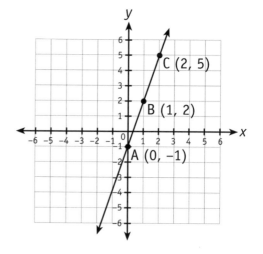

Table of Values

x value	$y = 3x - 1$	x	y
$x = 0$	$y = 3(0) - 1 = -1$	0	-1
$x = 1$	$y = 3(1) - 1 = 2$	1	2
$x = 2$	$y = 3(2) - 1 = 5$	2	5

STEP 2 Plot the points A = (0, -1) B = (1, 2) C = (2, 5). Connect them with a line that extends to the edges of the grid.

Graph each equation. As a first step, complete the Table of Values. Then find the x- and y-intercepts and slope for each graphed line.

1. $y = x - 2$

Table of Values

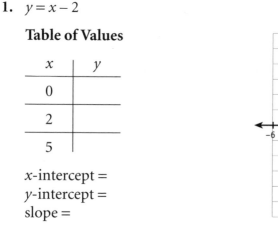

x	y
0	
2	
5	

x-intercept =
y-intercept =
slope =

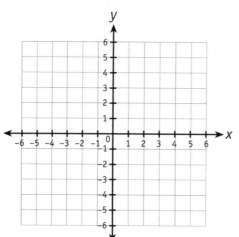

2. $y = x + 2$
Choose your own values for x.

Table of Values

x	y

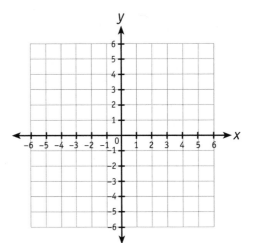

x-intercept =

y-intercept =

slope =

3. $y = -2x - 4$
Choose your own values for x.

Table of Values

x	y

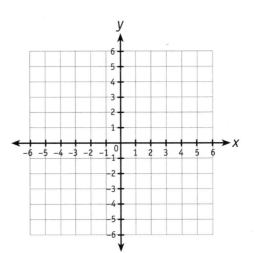

x-intercept =

y-intercept =

slope =

4. $y = \frac{3}{4}x - 1$

Table of Values

x	y
0	
2	
4	

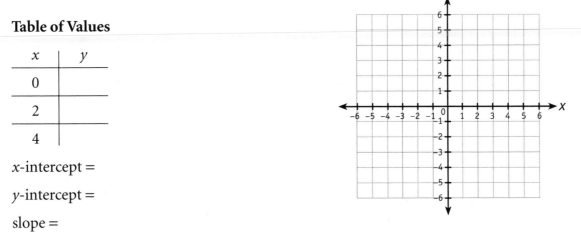

x-intercept =

y-intercept =

slope =

Graphing Circumference

The formula for the circumference of a circle, $C = 2\pi r$, is a linear equation containing the variables C and r. The number π is *not* a variable. The approximate value of π is 3.14.

Use the grid at the right to answer the problems below.

1. Graph the equation $C = 2\pi r$ for positive values for r and C. For the Table of Values, choose four values for r from among those shown on the grid.

Table of Values

r	C

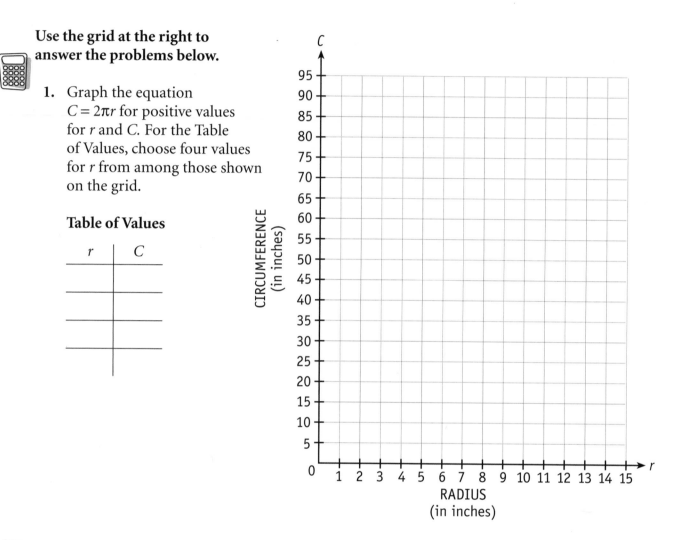

2. Estimate the circumference of a circle that has a radius of $6\frac{1}{2}$ inches.

3. Estimate the radius of a circle that has a circumference of 48 inches.

4. Estimate the slope of the graphed line $C = 2\pi r$.

$$\text{slope} = \frac{\text{change in } C}{\text{change in } r}$$

Graphing Temperature

The formula that relates Celsius (°C) to Fahrenheit (°F) temperature is $°C = \frac{5}{9}(°F - 32°)$. This formula is a linear equation containing the variables °F and °C. When finding values, remember to find the difference within parentheses before multiplying by $\frac{5}{9}$.

Use the grid at the right to answer the problems below.

1. Graph the equation $°C = \frac{5}{9}(°F - 32°)$ for positive values of °F.

    For the Table of Values, choose four values for °F from among those shown on the grid.

Table of Values

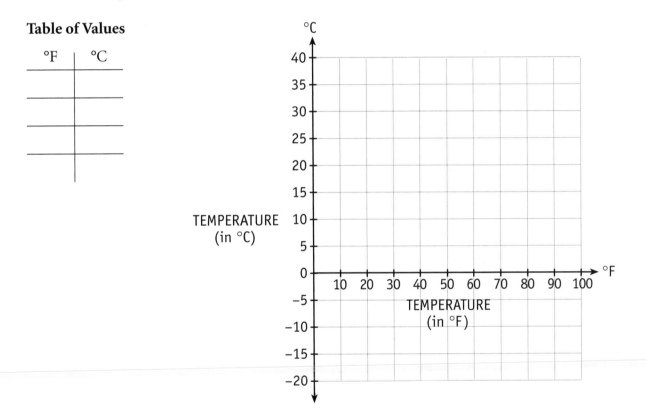

°F	°C

2. Estimate the temperature in °C when a Fahrenheit thermometer reads 65°F.

3. At about what temperature does the graphed line cross the horizontal (°F) axis?

4. Estimate the slope of the graphed line $°C = \frac{5}{9}(°F - 32°)$.

    $$\text{slope} = \frac{\text{change in °C}}{\text{change in °F}}$$

Graphing a System of Equations

On page 94, you learned that two linear equations with a common solution is called a system of equations. One way to find a common solution is to subtract one equation from the other.

A second way to find a common solution is to graph each equation on the same graph. The point at which the two lines cross is the common solution.

EXAMPLE   By graphing, find the common solution to these equations.
Equation 1: $y = 2x + 3$
Equation 2: $y = x + 2$

STEP 1   Make a Table of Values for each equation.

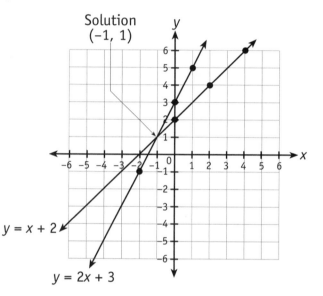

Table of Values
for $y = 2x + 3$

x	y
−2	−1
0	3
1	5

Table of Values
for $y = x + 2$

x	y
0	2
2	4
4	6

STEP 2   Graph each equation.

STEP 3   Identify the point at which the lines intersect. This point is (−1, 1).

ANSWER: Common solution is (−1, 1).

Find a common solution for the two equations by graphing. Complete the Table of Values. Check your answers by subtracting one equation from the other.

1.   Equation 1: $y = 2x + 1$
Equation 2: $y = -x + 4$

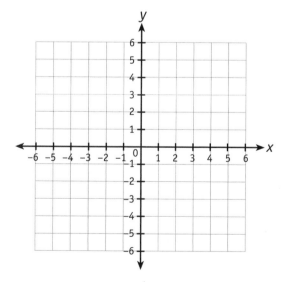

Table of Values
for $y = 2x + 1$

x	y
−2	
0	
2	

Table of Values
for $y = -x + 4$

x	y
0	
2	
4	

Common solution: (,   )

2. Equation 1: $y = 2x + 3$
Equation 2: $y = x + 1$

Table of Values
for $y = 2x + 3$

x	y
-1	
0	
1	

Table of Values
for $y = x + 1$

x	y
0	
2	
4	

Common solution: (,)

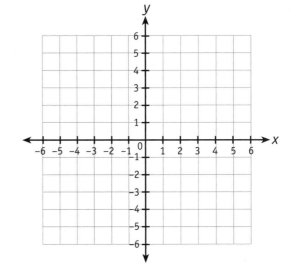

3. Equation 1: $y = 3x + 1$
Equation 2: $y = -x + 5$

Table of Values
for $y = 3x + 1$

x	y
-1	
0	
1	

Table of Values
for $y = -x + 5$

x	y
-1	
0	
6	

Common solution: (,)

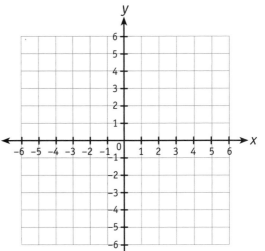

Graphing a Quadratic Equation

When a quadratic equation such as $y = x^2 - 3$ is graphed, the resulting graph is a curved line called a **parabola.**

To graph a quadratic equation, follow these steps.

STEP 1 Choose both negative and positive values for x. Then find the matching values for y. Write these in a table.

STEP 2 Plot each point on the grid, and connect them with a curved line. Extend the curve to the edges of the grid.

EXAMPLE 1 Graph the equation $y = x^2 - 3$.

STEP 1 Let $x = -3, -2, -1, 0, 1, 2,$ and 3. Solve the equation $y = x^2 - 3$ for these values. Make a Table of Values.

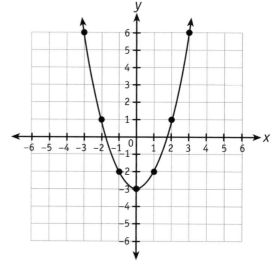

Table of Values

x value	$y = x^2 - 3$	x	y
$x = -3$	$y = (-3)^2 - 3 = 6$	-3	6
$x = -2$	$y = (-2)^2 - 3 = 1$	-2	1
$x = -1$	$y = (-1)^2 - 3 = -2$	-1	-2
$x = 0$	$y = (0)^2 - 3 = -3$	0	-3
$x = 1$	$y = (1)^2 - 3 = -2$	1	-2
$x = 2$	$y = (2)^2 - 3 = 1$	2	1
$x = 3$	$y = (3)^2 - 3 = 6$	3	6

STEP 2 Plot each point and connect them with a smoothly curving line. See the completed graph.

EXAMPLE 2 Use the graph from Example 1 to estimate x when $y = 4$.

STEP 1 Locate +4 on the y-axis.

STEP 2 From +4 on the y-axis, read the left and right points on the curve. Read down to the x-axis for the two x values.

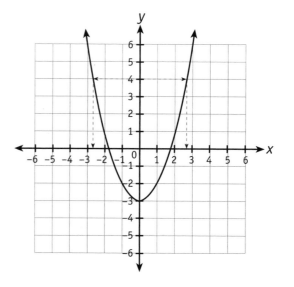

ANSWER: x can have either of two values:
$x \approx -2.7$ and $x \approx 2.7$

These values are the two approximate square roots of 7. Do you know why?

In the graphed equation, $x^2 = 7$ when $y = 4$.

For each problem, graph the equation. As a first step, complete the
Table of Values. (The first value is done for you.) Then use the graph to
estimate the two values for x as indicated.

1. **a.** $y = x^2 - 5$

   Table of Values

x	y
-4	11
-3	
-2	
0	
2	
3	
4	

   b. Estimate the two values of x
   when $y = 6$.

2. **a.** $y = 2x^2 - 6$

   Table of Values

x	y
-3	12
-2	
-1	
0	
1	
2	
3	

   b. Estimate the two values of
   x when $y = 5$.

Graphing Equations Review

Solve the problems below. When you finish, check your answers at the back of the book. Then correct any errors.

For problems 1 and 2, refer to the grid at the right.

1. On the grid, label the following.

    a. the x-axis with an x

    b. the y-axis with a y

    c. the origin with $(0, 0)$

    d. the x-coordinates from -6 to $+6$

    e. the y-coordinates from -6 to $+6$

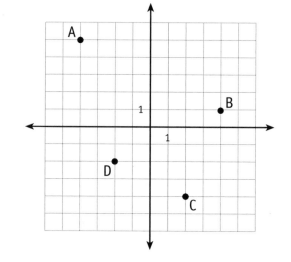

2. Write the coordinates of each point.

    Point A = (,   )

    Point B = (,   )

    Point C = (,   )

    Point D = (,   )

For problem 3, refer to the grid at the right.

3. **a.** Graph the line containing the points $(-6, -4)$, $(-3, -3)$, and $(3, -1)$.

    b. Identify the x- and y-intercepts of the line you drew.

    x-intercept = (,   )

    y-intercept = (,   )

    c. Find the slope of your graphed line.

    slope =

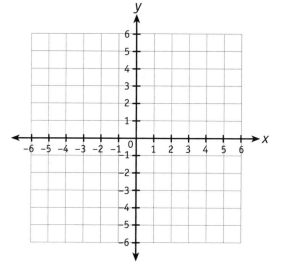

Find the slope of the line passing through each set of points.

4. **a.** $(0, 0)$ and $(2, 4)$ **b.** $(1, 0)$ and $(2, 5)$ **c.** $(2, 3)$ and $(4, 3)$

5. Graph the equation $y = 2x - 4$. Choose your own values for x. Find the x- and y-intercepts and the slope of the graphed line.

Table of Values

x	y

x-intercept =

y-intercept =

slope =

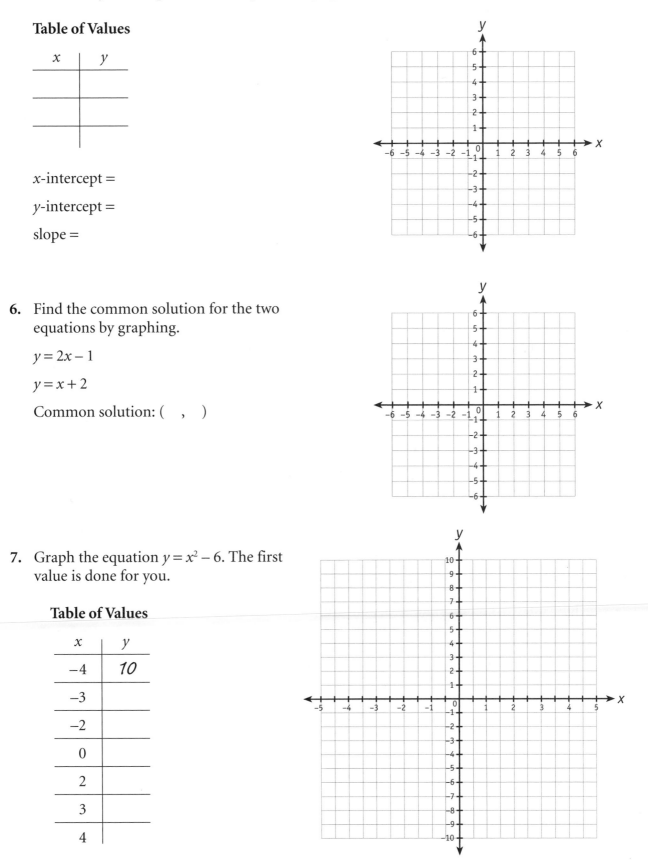

6. Find the common solution for the two equations by graphing.

$y = 2x - 1$

$y = x + 2$

Common solution: (,   )

7. Graph the equation $y = x^2 - 6$. The first value is done for you.

Table of Values

x	y
−4	10
−3	
−2	
0	
2	
3	
4	

INEQUALITIES

What Is an Inequality?

In an earlier chapter, you learned that an equation is a statement that two quantities are equal. In this chapter, you will learn about inequalities. An **inequality** is a statement that two quantities are *not* equal.

Five comparison symbols are used with inequalities.

Symbol	Meaning	Example	Read as
$<$	is less than	$b < 9$	b is less than 9
$>$	is greater than	$n > 16$	n is greater than 16
$\leq$	is less than *or* equal to	$x \leq 5$	x is less than *or* equal to 5
$\geq$	is greater than *or* equal to	$s \geq -4$	s is greater than *or* equal to -4
$\neq$	is not equal to	$c \neq 7$	c is not equal to 7

In an inequality, the variable can be *any value* for which the inequality is true.

- $b < 9$ b can be any value less than 9.

  Example values: $b = -2.5$, $b = 0$, or $b = 8$

- $n > 16$ n can be any value greater than 16.

  Example values: $n = 16.3$, $n = 17$, or $n = 31\frac{1}{4}$

- $a \leq 7$ a can be any value less than *or* equal to 7.

  Example values: $a = -3.5$, $a = 1$, or $a = 7$

- $r \geq -2$ r can be any value greater than *or* equal to -2.

  Example values: $r = -2$, $r = 0$, or $r = 2\frac{1}{2}$

> **Did You Know . . . ?**
> In an inequality, the allowed values are whole numbers, decimals, and fractions.

**For each inequality, circle any allowed value for the variable.
Two problems are done as examples.**

1. **a.** $m < 7$ ⓸(-3)(0)(5)   7   9   14   **b.** $n < 10$ -2 0   3   7   10   15

2. **a.** $s > 6$ -6 -2 6   8   10   19   **b.** $y > 3$ -3 0   3   6   11   23

3. **a.** $n \leq 12$ (-8)(4)   (12)   13   19   25   **b.** $p \leq 6$ -5 0   3   5   6   27
   [n can be any number less than or equal to 12.]   [p can be any number less than or equal to 6.]

4. **a.** $x \geq 14$ -6 0   2   17   29   41   **b.** $z \geq -4$ -9 -5 -4 -3 0   4
   [x can be any number greater than or equal to 14.]   [z can be any number greater than or equal to -4.]

Graphing an Inequality

The values for which an inequality is true can be graphed on a number line.

- A solid circle ● means that the number *is* an allowed value.
- An open circle ○ means that a number is *not* an allowed value.
- A solid line is drawn through all allowed values.

__EXAMPLE 1__ Inequality: $n > -5$

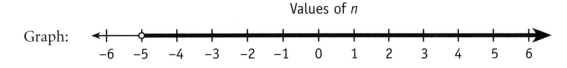

Values of *n*

Graph:

Meaning: *n* is greater than –5 [*n cannot* have the value –5, indicated by ○.]

__EXAMPLE 2__ Inequality: $r \geq 0$

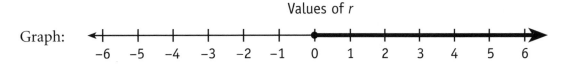

Values of *r*

Graph:

Meaning: *r* is greater than or equal to 0. [*r can* have the value 0, indicated by ●.]

__EXAMPLE 3__ Inequality: $y \leq -2$

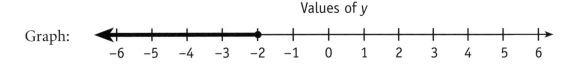

Values of *y*

Graph:

Meaning: *y* is less than or equal to –2. [*y can* have the value –2, indicated by ●.]

Circle the letter for the inequality graphed on each number line.

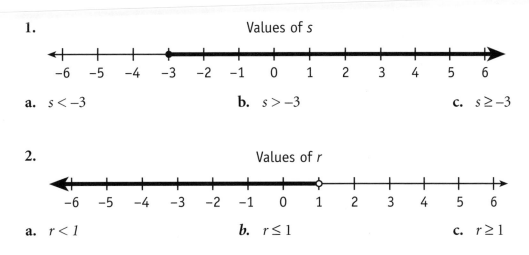

1.

Values of *s*

a. $s < -3$ b. $s > -3$ c. $s \geq -3$

2.

Values of *r*

a. $r < 1$ b. $r \leq 1$ c. $r \geq 1$

Write the inequality graphed on each number line.

3.

Values of *s*

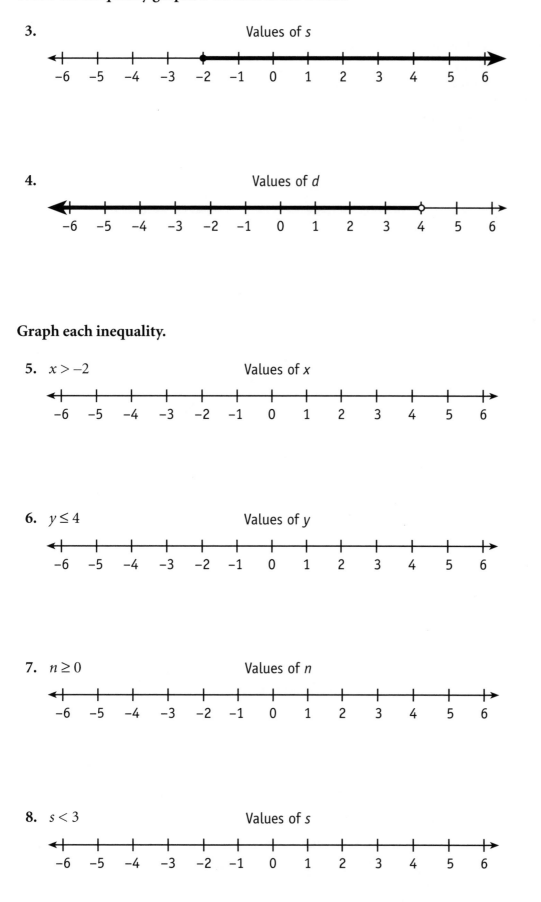

4.

Values of *d*

Graph each inequality.

5. $x > -2$

Values of *x*

6. $y \leq 4$

Values of *y*

7. $n \geq 0$

Values of *n*

8. $s < 3$

Values of *s*

Solving an Inequality

In an inequality, the variable may not be standing alone and the allowed values are not easy to see. For example, look at the inequality $2x + 4 \geq 5$. To find the allowed values of x, you solve this inequality in a way similar to the way you solve an equation.

To solve an inequality is to write it so that the variable stands alone.

To get the variable standing alone, perform the same operations on each side of the inequality sign.

- Add (subtract) the same number to (from) each side. (See Example 1.)
- Multiply or divide each side by the same positive number. (See Example 2.)
- Use two or more operations, starting with addition or subtraction. (See Examples 3 and 4.)

EXAMPLE 1 Solve for x: $x - 5 \leq 2$

    STEP 1 Add 5 to each side of the inequality.

        $x - 5 + 5 \leq 2 + 5$

    STEP 2 Simplify each side of the inequality.

ANSWER: $x \leq 7$

EXAMPLE 2 Solve for y: $4y \geq 16$

    STEP 1 Divide each side of the inequality by 4.

        $\frac{4y}{4} \geq \frac{16}{4}$

    STEP 2 Simplify each side of the inequality.

ANSWER: $y \geq 4$

EXAMPLE 3 Solve for y: $5y - 3 \leq 27$

    STEP 1 Add 3 to each side of the inequality.

        $5y - 3 + 3 \leq 27 + 3$

    STEP 2 Simplify each side of the inequality.

        $5y \leq 30$

    STEP 3 Divide each side of the inequality by 5.

        $\frac{5y}{5} \leq \frac{30}{5}$

    STEP 4 Simplify each side of the inequality.

ANSWER: $y \leq 6$

EXAMPLE 4 Solve for n: $\frac{n}{2} + 6 \leq 13$

    STEP 1 Subtract 6 from each side of the inequality.

        $\frac{n}{2} + 6 - 6 \leq 13 - 6$

    STEP 2 Simplify each side of the inequality.

        $\frac{n}{2} \leq 7$

    STEP 3 Multiply each side of the inequality by 2.

        $2\left(\frac{n}{2}\right) \leq 2(7)$

    STEP 4 Simplify each side of the inequality.

ANSWER: $n \leq 14$

Solve each inequality.

1. $2x + 6 \geq 10$ $3n - 4 \geq 11$ $5z + 5 > 25$

2. $6w + 2 \leq 14$ $4x - 1 < -9$ $2y + 4 \leq 0$

3. $\frac{n}{2} + 3 > 9$ $\frac{x}{5} - 6 \geq 2$ $\frac{c}{4} - 3 < -2$

4. $\frac{2}{3}x + 1 \leq 5$ $\frac{3}{4}n + 3 < -3$ $\frac{5}{8}y - 4 \leq 21$

· ·

   If you multiply or divide each side of an inequality by –1, you need to reverse the inequality sign.

EXAMPLE 5   Solve for y: $-y \leq 6$

    STEP 1   Multiply each side of the inequality by –1 and change the inequality sign from ≤ to ≥.

          $-1(-y) \geq -1(6)$

    STEP 2   Simplify both sides of the inequality.

ANSWER: $y \geq -6$

EXAMPLE 6   Solve for x: $3 - x > 5$

    STEP 1   Subtract 3 from each side of the inequality and simplify.

          $3 - 3 - x > 5 - 3$
          $-x > 2$

    STEP 2   Multiply each side of the inequality by –1, change > to < and simplify.

          $-1(-x) < -1(2)$
          $x < -2$

ANSWER: $x < -2$

· ·

5. $-x \geq 8$ $4 - n < 10$ $6 - 2x > 10$

Picturing an Inequality

An inequality can be pictured on a line drawing.

EXAMPLE   On the drawing at the right, the sum of the lengths of the labeled segments of the divided line is less than the total length 19 of the undivided line.

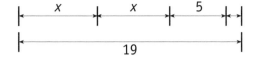

$$x + x + 5 < 19$$
or **2x + 5 < 19**

The possible values of x are found by solving the inequality.

Solve: $2x + 5 < 19$
$2x + 5 - 5 < 19 - 5$
$2x < 14$
$\frac{2x}{2} < \frac{14}{2}$

Because x is a distance, you know that the allowed values of x are those numbers that are greater than 0 but less than 7.

ANSWER: x < 7

Write and solve the inequality for each line drawing.

1.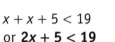

Inequality:

Solution:

2.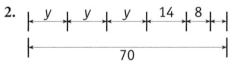

Inequality:

Solution:

3.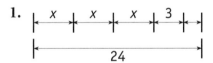

Inequality:

Solution:

4.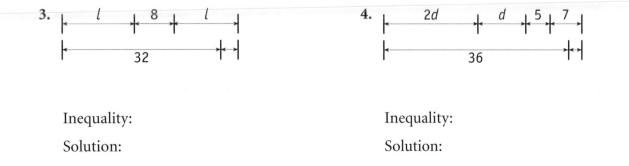

Inequality:

Solution:

Inequalities: Word Problem Skills

Circle the correct inequality and solve for the allowed values of the variable.

1. If 6 is added to a number *n*, the sum is greater than 10. Which inequality tells the possible values of *n*?

    a. $n + 6 > 4$
    b. $n - 10 > 6$
    c. $n + 6 > 10$

    Solution:

2. If 8 is subtracted from 3 times a number *m*, the difference is less than or equal to 4. Which inequality tells the possible values of *m*?

    a. $3m - 8 \leq 4$
    b. $8 - 3m \leq 4$
    c. $3m - 8 \geq 4$

    Solution:

Solve.

3. Jon will be paid $35.00 for working Saturday. He will also be paid an extra $1.50 for each large poster he sells. Jon's goal is to earn at least $85.00 on Saturday.

    a. Write an inequality that tells how many posters (*n*) Jon must sell on Saturday in order to reach or exceed his goal.

    b. What is the least number of posters that Jon must sell in order to earn $85.00 or more?

4. The perimeter of △ABC is greater than 75 yards.

    a. Write an inequality that describes the possible values for the length (*c*) of the third side.

    b. To the nearest yard, what is the smallest value *c* may have?

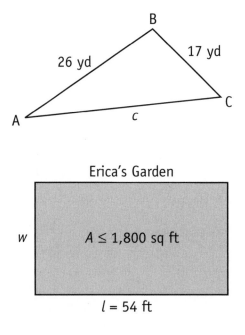

5. Erica is designing a garden to be 54 feet long and no larger than 1,800 square feet in area.

    a. Write an inequality that describes the range of possible values for the width (*w*) of the garden.

    b. To the nearest foot, what is the largest value *w* may have?

Erica's Garden

w $A \leq 1{,}800$ sq ft

$l = 54$ ft

Two-Part Inequalities

A two-part inequality contains two comparison symbols that are used to describe *a range of allowed values.* The two types of two-part inequalities are ***and* inequalities** and ***or* inequalities.**

And Inequalities

Inequalities whose allowed values run *between* two numbers are called *and inequalities.* The allowed values of an *and inequality* can be graphed as a single section of a number line.

Examples	Meaning	Graphed Values
$2 < n < 6$	n is greater than 2 *and* n is less than 6.	
$-3 \leq m < 4$	m is greater than or equal to -3 *and* m is less than 4.	
$-5 \leq y \leq 5$	y is greater than or equal to -5 *and* y is less than or equal to 5.	

Or Inequalities

In an *or inequality,* allowed values are graphed as separated sections of a number line.

Examples	Meaning	Graphed Values
$s < 2$ or $s \geq 4$	s is less than 2 *or* s is greater than or equal to 4.	
$x \leq -1$ or $x > 3$	x is less than or equal to -1 *or* x is greater than 3.	

Write the meaning of each inequality.

1. $-5 < n < 8$ _____

2. $-1 \leq x < 11$ _____

3. $n \leq 0$ *or* $n > 7$ _____

4. $p < -3$ or $p \geq 5$ _____

For each inequality, three values are given. Circle Yes if the value is a possible value for the inequality, No if it is not.

5. $-2 < y < 5$ **a.** $y = 3$ Yes No   **b.** $y = -2$ Yes No   **c.** $y = 0$ Yes No

6. $-5 \le x < 0$ **a.** $x = 0.6$ Yes No   **b.** $x = 5$ Yes No   **c.** $x = -5$ Yes No

7. $2 < n \le 8$ **a.** $n = -3$ Yes No   **b.** $n = 5\frac{1}{3}$ Yes No   **c.** $n = 8$ Yes No

8. $y \le 3$ or $y > 8$ **a.** $y = -1.5$ Yes No   **b.** $y = 0$ Yes No   **c.** $y = 8$ Yes No

9. $x < -2$ or $x \ge 3$ **a.** $x = -4$ Yes No   **b.** $x = 1$ Yes No   **c.** $x = 3$ Yes No

10. $n \le 1$ or $n \ge 5$ **a.** $n = -1$ Yes No   **b.** $n = 0$ Yes No   **c.** $n = 5\frac{3}{4}$ Yes No

Write the inequality graphed on each number line.

11.

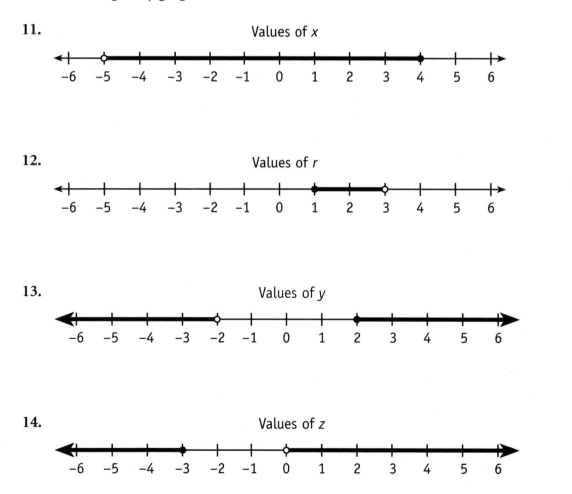

Values of x

12.

Values of r

13.

Values of y

14.

Values of z

Graph each inequality.

15. $-3 \leq n < 4$ Values of n

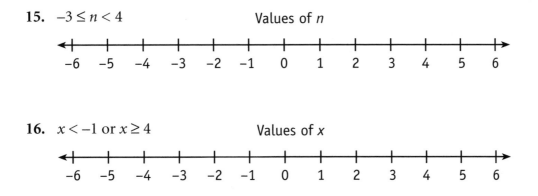

16. $x < -1$ or $x \geq 4$ Values of x

Solve.

17. During the year, Mindy's weight varies from 132 pounds to 138 pounds. She tends to gain weight during the winter and reduce during the summer. Write an inequality that describes the range of Mindy's weight (w).

18. Each day, 5% to 10% of the 2,000 students who attend Hoover High School are absent. Write an inequality that describes the number of students (n) who are absent on an average day at Hoover High School.

19. The perimeter of ΔDEF is greater than 102 inches but less than 119 inches.

    a. Write an inequality that describes the range of possible values for the length (c) of the third side.

    b. If c is a whole number of inches, what is the smallest value c may be?

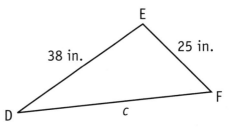

20. Geraldine conducted a survey. She contacted only people who have been driving for less than 1 year or for more than 5 years. Write an inequality that gives the range of values for the length of time (t) that the people in the survey have been driving.

Inequalities Review

Solve each problem below. When you finish, check your answers at the back of the book. Then correct any errors.

Write the meaning of each inequality.

1. $x < 6$ _____

2. $n > 9$ _____

3. $y \geq -5$ _____

4. $s \leq 4$ _____

Write the inequality graphed on each number line.

5.
Values of p

6. Values of r

7.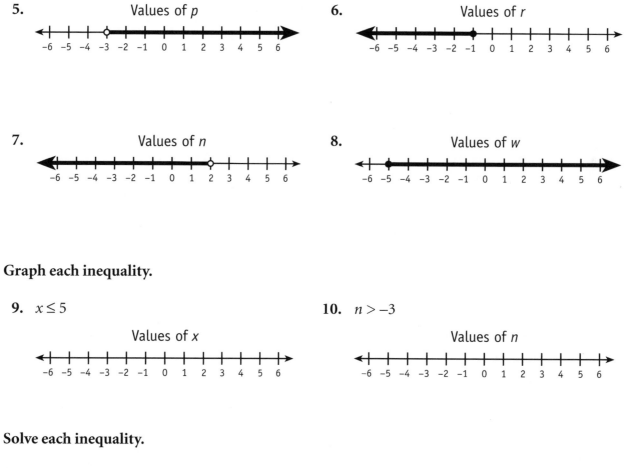
Values of n

8. Values of w

Graph each inequality.

9. $x \leq 5$

10. $n > -3$

Values of x

Values of n

Solve each inequality.

11. $3x + 4 \leq 10$

12. $2n - 5 \geq 15$

Write and solve the inequality for each line drawing.

13.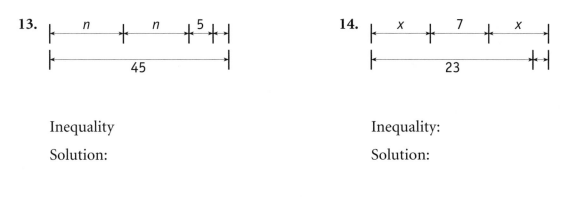

    Inequality

    Solution:

14.

    Inequality:

    Solution:

Write the meaning of each inequality.

15. $-2 < x < 7$ _____

16. $0 \le y < 6$ _____

17. $n \le -2 \text{ or } n \ge 5$ _____

18. $r < 2 \text{ or } r > 7$ _____

Write the inequality graphed on each number line.

19.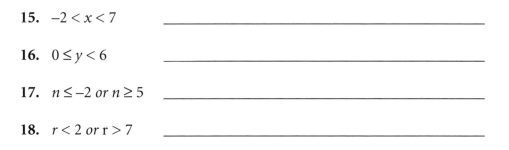

Values of y

20.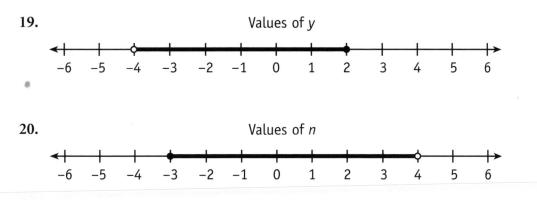

Values of n

Solve.

21. The perimeter of $\triangle ABC$ is less than 94 centimeters but greater than 85 centimeters.

    a. Write an inequality that describes the range of possible values for the length of side BC.

    b. If side BC is a whole number of centimeters, what is the shortest length side BC may be?

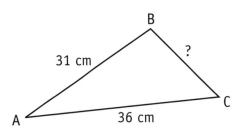

POLYNOMIALS

Naming Polynomials

A **polynomial** is an algebraic expression that contains one or more terms combined by addition. Each term is the product of a number (coefficient) and one or more variables. Each variable may have an exponent. A number standing alone is also a term.

The three most common polynomials are **monomials, binomials,** and **trinomials.**

- A monomial has one term.
- A binomial has two terms.
- A trinomial has three terms.

Monomials	Binomials	Trinomials
z	$x - 3$	$x + y + z$
$-2x$	$3y + 9$	$5r - 3s + 6t$
$3a^2$	$-5c^2 + 2c$	$4y^2 - 5y + 7$
$-6x^2y^2$	$4a^2b^2 + 2ab$	$7y^3 - 10y^2 + 9y$

To identify a term, change a subtraction sign to an addition sign and change the sign of the coefficient of the term. For example, in the binomial $x - 3$ the terms are x and -3. In the trinomial $7y^3 - 10y^2 + 9y$, the terms are $7y^3$, $-10y^2$, and $9y$.

Name each polynomial as a monomial, binomial, or trinomial.

1. $5x - 6$ _____

2. $4x + 6y - z$ _____

3. $-y$ _____

4. $-5y^2 + 3y$ _____

5. $r^2 + s^2 - t^2$ _____

6. $7x^2y^2$ _____

Identify the terms in each polynomial. The first problem in each row is done as an example.

7. $-12x + 7$ $9d + 6$ $-3a^2 + 5a$ $4y + \frac{1}{2}$
   $-12x$ and 7

8. $9a - 2b$ $-6d - 8e$ $8x^2 - 3y^2$ $\frac{2}{3}z^2 - 7z$
   $9a$ and $-2b$

9. $2x^2 + 7x - 4$ $3y^2 - 5y - 8$ $8x - 4y + 13$
   $2x^2$, $7x$, and -4

Recognizing Like Terms

As mentioned on the previous page, a term in a polynomial can be a number standing alone, or it can be the product of a coefficient times one or more variables with exponents. The variables with exponents are called the **variable part** of the term.

EXAMPLE

$$-2x^2y$$

coefficient —↑ ↑— variable part

Like terms are terms that have identical variable parts. Like terms differ only in coefficients. For example, $8x^2y$ and $-2x^2y$ are like terms because the variable part (x^2y) is the same. All numbers standing alone are like terms. For example, -4 and 9 are like terms.

Terms with different variable parts are called **unlike terms**.

Like Terms	Unlike Terms
5 and -7	x and y
$-3y$ and $4y$	$-r$ and s
$5x^2$ and $2x^2$	$4x^2$ and $-2z$
$4a^2b$ and $-a^2b$	a^2b and b^2a

Identify the coefficient (C) and the variable part (VP) of each term. The first problem in each row is done as an example.

	C	VP		C	VP		C	VP
1. $-4x$	-4	x	$3a$	___	___	$-5z$	___	___
2. y^2	1	y^2	z^2	___	___	$-a^2$	___	___
3. $-9x^2y$	-9	x^2y	$-4c^2d$	___	___	$7uv^2$	___	___
4. $3a^2b^2$	3	a^2b^2	$8x^2y^2$	___	___	$-7ab^3$	___	___

Circle the pair of like terms in each group.

5. $x, x^2, -5x$ $y^3, 3y, 2y^3$ $a^2b, 2a^2b, b^2a$

6. $a^2, -4a^2, 2a, 5a^4$ $-2x, 2xy^2, -2x^2y, xy^2$ $-3x, 3x^2, x^3, 4x^3$

7. $2c, c^2, -c^3, -5c^2$ $3rs, -rs^2, 3r^2, 3rs^2$ $-4x^3y, xy^4, 3x^3y^3, 2x^3y$

Adding Polynomials

To add polynomials, add the coefficients of like terms. Follow the rules for adding signed numbers.

EXAMPLE 1   Add: $3x$, $2x$, and $4y$

$3x + 2x + 4y = \mathbf{5x + 4y}$
$(3 + 2 = 5)$
x and y are unlike terms.

EXAMPLE 2   Add: $-5a^2b$ and a^2b.

$-5a^2b + a^2b = \mathbf{-4a^2b}$
$(-5 + 1 = -4)$
The coefficient of a^2b is 1.

EXAMPLE 3   Add: $(6x - 9y) + (-5x + 4y)$

STEP 1   Remove the parentheses from each binomial.

$6x - 9y + (-5x) + 4y$

STEP 2   Combine like terms.

x terms: $6x + (-5x) = \mathbf{x}$ $(6 + -5 = 1)$
y terms: $-9y + 4y = \mathbf{-5y}$ $(-9 + 4 = -5)$

ANSWER: $x + -5y$ or $x - 5y$

(**Remember:** An added negative number can be written as a subtracted positive number.)
The answer can be written either way. However, $x - 5y$ is the usual way to write this answer.

Add. The first problem in each row is done as an example.

Adding Monomials

1. $6y + 3y = \mathbf{9y}$ $5r + 2r =$ $12s + 7s =$ $10x + 5x =$

2. $4z + z = \mathbf{5z}$ $7a + a =$ $2c + c =$ $11n + n =$

3. $8x + (-3x) = \mathbf{5x}$ $9y + (-2y) =$ $15r + (-4r) =$ $7y + (-7y) =$

4. $2x^2 + (-4x^2) = \mathbf{-2x^2}$ $6y^2 + (-y^2) =$ $7z^3 + (-2z^3) =$ $12r^3 + (-4r^3) =$

Adding Binomials and Monomials

5. $(3y + 5) + 7 = 3y + 12$ $(4x + 9) + 4 =$ $(7z + 6) + (-3) =$

6. $(4z + 2) + -5z = -z + 2$ $(7y + -8) + y =$ $(12s + 4) + 4s =$

7. $(7n^2 + n) + 9n^2 = 16n^2 + n$ $(6x^2 + 5x) + 3x =$ $(3r^2 + r) + (-2r^2) =$

Adding Binomials and Binomials

8. $(2z + 4) + (-3z + 5) = -z + 9$ 9. $(5a + 3b) + (-2a - 8b) = 3a - 5b$

   $(4x + 6) + (-2x + 5) =$ $(2x + 9y) + (-x - 5y) =$

   $(5n + 1) + (-n + 7) =$ $(8r + 3s) + (-3r - 4s) =$

10. $(4z^2 + 6) + (-z^2 - 4) = 3z^2 + 2$ 11. $(5a^2 + 3b) + (a^2 - 5b) = 6a^2 - 2b$

   $(6n^2 + 5) + (-3n^2 - 2) =$ $(6r^2 + 5s) + (2r^2 - 8s) =$

   $(2y^2 + 9y) + (-y^2 - 5y) =$ $(9m^2 + n) + (4m^2 - 3n) =$

Write the perimeter of each figure as a trinomial.

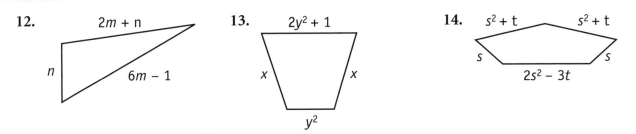

12. $2m + n$ $6m - 1$ n

13. $2y^2 + 1$ x x y^2

14. $s^2 + t$ $s^2 + t$ s s $2s^2 - 3t$

Subtracting Polynomials

To subtract polynomials, subtract the coefficients of like terms. To subtract coefficients, follow the rules for subtracting signed numbers.
- Change the subtraction sign (−) to an addition sign (+).
- Change the sign of each term being subtracted.

EXAMPLE 1 Subtract: $9y - 4y$

Subtract coefficients of the y terms.

$9 - 4 = \mathbf{5}$ or $9 + (-4) = \mathbf{5}$

ANSWER: 5y

EXAMPLE 2 Subtract: $5c^2 - (-6c^2)$

Subtract coefficients of the c^2 terms.

$5 - (-6) = 5 + 6 = \mathbf{11}$

ANSWER: 11c²

(**Remember:** To subtract, change − to + and change −6 to 6.)

When a subtraction sign precedes a binomial in parentheses,
- remove the parentheses
- subtract each term that was within the parentheses

EXAMPLE 3 Subtract: $(5x - y) - (3x + 4y)$

STEP 1 Remove the parentheses from each binomial.

$(5x - y) = 5x - y$
$-(3x + 4y) = -3x - 4y$

Subtract each term.

STEP 2 Combine like terms.

x terms: $5x - 3x = \mathbf{2x}$
y terms: $-y - 4y = \mathbf{-5y}$

ANSWER: 2x − 5y

Subtract. The first problem in each row is done as an example.

1. $5x - 2x = \mathbf{3x}$ $7z - 2z =$ $10r - 6r =$ $15x - 8x =$

2. $6z - (-3z) = \mathbf{6z + 3z}$ $9b - (-b) =$ $6d - (-2d) =$ $21n - (-17n) =$
       $= \mathbf{9z}$

3. $(7x + 2y) - 5x = 7x - 5x + 2y = 2x + 2y$ 4. $(3x - 3y) - 5y = 3x - 3y - 5y = 3x - 8y$

   $(8m + 3n) - 2m =$ $(9a + 5b) - 6a =$

   $(5r - 6s) - 7s =$ $(12c - 3d) - 7d =$

5. $(8a - 3b) - (2a + b) = 8a - 2a - 3b - b = 6a - 4b$

   $(9x - 2y) - (6x + 2y) =$

   $(12m + 3n) - (5m + 5n) =$

6. $(5x^2 + 4x - 7) - (2x^2 + 2) = 5x^2 - 2x^2 + 4x - 7 - 2 = 3x^2 + 4x - 9$

   $(9a^2 + 2a - 8) - (3a^2 + 1) =$

   $(8n^2 + 7n - 2) - (n^2 + 9n + 2) =$

Write the unlabeled length as a polynomial.

7.

   2a + 3b   ?

   8a − 2b

8. 7x² − 3x   ?

   12x² − 5x + 7

Solve.

9. The perimeter of the figure at the right is given as the polynomial $8n^2 + 4n + 5$. Write the length of the unlabeled side as a polynomial.

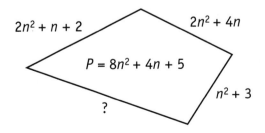

$2n^2 + n + 2$ $2n^2 + 4n$

$P = 8n^2 + 4n + 5$

$n^2 + 3$

?

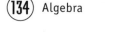

Multiplying Monomials

To multiply monomials, follow these two steps.

STEP 1 Multiply the coefficients according to the rules for multiplying signed numbers.

STEP 2 Multiply the variables by adding the exponents of each like variable. Write the variables in alphabetical order.

(**Note:** To review the rules for multiplying like bases by adding exponents, see page 28.)

Rule for multiplying powers of like variables:
Add the exponents and keep the same variable.

__EXAMPLE 1__ Multiply: $3x^2$ by $-4x$

    STEP 1 Multiply the coefficients.

        $3(-4) = \mathbf{-12}$

    STEP 2 Add the exponents of the x variables.

        $x^2x = x^2x^1 = x^{2+1} = \mathbf{x^3}$

ANSWER: $-12x^3$

__EXAMPLE 2__ Multiply: $-2a^2b(-3ab^3)$

    STEP 1 Multiply the coefficients.

        $-2(-3) = \mathbf{6}$

    STEP 2 Add the exponents of the a variables and the b variables.

        $a^2a = a^2a^1 = a^{2+1} = a^3$
        $bb^3 = b^1b^3 = b^{1+3} = b^4$

ANSWER: $6a^3b^4$

(**Remember:** If no exponent is written, it is understood to be 1. In Example 1, $x = x^1$. In Example 2, $a = a^1$ and $b = b^1$.)

Multiply. The first problem in each row is done as an example.

1. $5y^2(3y^3) = 15y^5$ $6x^4(2x^2) =$ $7n(3n^4) =$ $8a^3(6a^4) =$

2. $5x^2 \cdot 5x^2 = 25x^4$ $6n^4 \cdot 6n^4 =$ $4z^6 \cdot 4z^6 =$ $-8x^3 \cdot -8x^3 =$

3. $3x^2(-2x) = -6x^3$ $-4c^5(5c^3) =$ $9b^4(-7b) =$ $-5s(9s^3) =$

4. $7x^2y(-xy) = -7x^3y^2$ $-5c^3d^2(cd^2) =$ $3ab^4(-4a^3b) =$ $-8r^2s(-7rs^3) =$

Multiplying a Polynomial by a Monomial

To multiply a binomial or trinomial by a monomial, multiply each term and combine the separate products. Remove parentheses around a binomial or trinomial by multiplying each term within parentheses by the monomial.

EXAMPLE 1 Multiply: $3x(2x^2 + 4)$

STEP 1 Multiply each term of the binomial by $3x$. Keep the $+$ sign that is between the terms in the binomial.

$$3x(2x^2 + 4) = 3x \cdot 2x^2 + 3x \cdot 4$$

STEP 2 Multiply each pair of terms. Add the exponents of x.

$$3x \cdot 2x^2 + 3x \cdot 4 = \mathbf{6x^3 + 12x}$$

ANSWER: $6x^3 + 12x$

EXAMPLE 2 Multiply: $2a(5a^2 - 4ab + b^2)$

STEP 1 Multiply each term of the trinomial by $2a$. Keep the signs that are between the terms in the trinomial.

$$2a(5a^2 - 4ab + b^2) =$$
$$2a \cdot 5a^2 - 2a \cdot 4ab + 2a \cdot b^2$$

STEP 2 Multiply each pair or terms. Add the exponents.

$$2a \cdot 5a^2 - 2a \cdot 4ab + 2a \cdot b^2 =$$
$$\mathbf{10a^3 - 8a^2b + 2ab^2}$$

ANSWER: $10a^3 - 8a^2b + 2ab^2$

Multiply. The first problem in each row is done as an example.

1. $4n(3n^2 + 2) = \mathit{12n^3 + 8n}$

   $5c(c + 7) =$

   $6a(2a^3 - 5) =$

2. $3x^2(2x^2 + 4x) = \mathit{6x^4 + 12x^3}$

   $2y(4y^2 - 3y) =$

   $3r(3r^3 + 8r^2) =$

3. $ab(3a^2b^2 - 6) = \mathit{3a^3b^3 - 6ab}$

   $cd(5c^3d + 3) =$

   $xy(x^2y^3 - 2) =$

4. $4mn(-2m^3n^2 - mn) = \mathit{-8m^4n^3 - 4m^2n^2}$

   $3a^2b(a^2b - 6ab) =$

   $5cd^2(3c^2d - 2cd) =$

5. $3x(4x^3 - 3xy + xy^2) = \mathit{12x^4 - 9x^2y + 3x^2y^2}$

   $4a(4a^2 + 6ab^2 - 7b^3) =$

   $6cd(c^3d^3 - 2c^2d^2 + cd) =$

Multiplying a Binomial by a Binomial

To multiply a binomial by a binomial follow these steps.

STEP 1 Multiply the first term of the first binomial by each term of the second binomial.

STEP 2 Multiply the second term of the first binomial by each term of the second binomial.

STEP 3 Combine the separate products.

<u>EXAMPLE 1</u> Multiply: $(2x + 3y)(4x - 5y)$

$$\overbrace{(2x + 3y)(4x - 5y)}^{\text{STEP 1}} = \overbrace{2x(4x - 5y)}^{\text{STEP 1}} + \underbrace{3y(4x - 5y)}_{\text{STEP 2}}$$

STEP 1 Multiply $2x$ by $(4x - 5y)$: $2x(4x - 5y) = 2x \cdot 4x - 2x \cdot -5y$

$$= 8x^2 - 10xy$$

STEP 2 Multiply $3y$ by $(4x - 5y)$: $3y(4x - 5y) = 3y \cdot 4x - 3y \cdot -5y$

$$= 12xy - 15y^2$$

STEP 3 Combine: $8x^2 - 10xy + 12xy - 15y^2 = 8x^2 + 2xy - 15y^2$

ANSWER: $8x^2 + 2xy - 15y^2$

Multiply. The first problem in each group is done as an example.

1. $(2y + 3)(4y + 2) = 2y \cdot 4y + 2y \cdot 2 + 3 \cdot 4y + 3 \cdot 2 = 8y^2 + 4y + 12y + 6 =$
   $8y^2 + 16y + 6$

   $(4a + 2)(2a + 1) =$

   $(5z + 3)(3z + 2) =$

2. $(4x + 3y)(2x - y) = 4x \cdot 2x - 4x \cdot y + 3y \cdot 2x - 3y \cdot y = 8x^2 - 4xy + 6xy - 3y^2 =$
   $8x^2 + 2xy - 3y^2$

   $(2a + 3b)(3a - 2b) =$

   $(3r + 2s)(r - 4s) =$

An interesting result occurs when you multiply the sum of two variables by the difference of the same variables:

$$(a + b)(a - b) = a^2 - ab + ab - b^2 = \boldsymbol{a^2 - b^2}$$

The product is the difference of the squares of the variables. In this product, a number can be substituted for either variable, a or b.

EXAMPLE 2 Multiply: $(x + 3)(x - 3)$

Multiply $x \times x$ and $3 \times (-3)$.

$(x + 3)(x - 3) = \boldsymbol{x^2 - 9}$

ANSWER: $\boldsymbol{x^2 - 9}$

EXAMPLE 3 Multiply: $(5 + n)(5 - n)$

Multiply 5×5 and $n \times (-n)$.

$(5 + n)(5 - n) = \boldsymbol{25 - n^2}$

ANSWER: $\boldsymbol{25 - n^2}$

3. $(a + 2)(a - 2) = a^2 - 4$

$(c + 5)(c - 5) =$

$(y + 8)(y - 8) =$

4. $(6 + x)(6 - x) = 36 - x^2$

$(9 + b)(9 - b) =$

$(4 + m)(4 - m) =$

Write the area of each rectangle as a polynomial. Formula is on page 209.

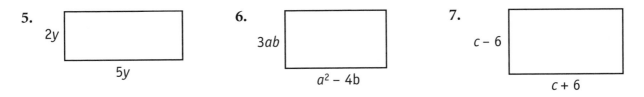

5. $2y$, $5y$

6. $3ab$, $a^2 - 4b$

7. $c - 6$, $c + 6$

Write the volume of each figure as a monomial or a binomial. Formulas are on page 209.

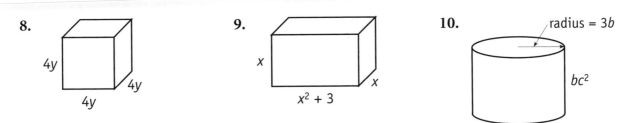

8. $4y$, $4y$, $4y$

9. x , x , $x^2 + 3$

10. radius = $3b$, bc^2

Dividing Monomials

To divide monomials, follow these two steps.

STEP 1 Divide the coefficients according to the rules for dividing signed numbers.

STEP 2 Divide the variables by subtracting the exponents of each like variable. Write the variables in alphabetical order.

(**Note:** To review the rules for dividing like bases by subtracting exponents, see page 29.)

> **Rule for dividing powers of like variables:**
> Subtract the exponents and keep the same variable.

EXAMPLE 1 Divide: $8y^4$ by $-4y$.

STEP 1 Divide the coefficients.

$$\frac{8}{-4} = -2$$

STEP 2 Subtract the exponents of the y variables.

$$\frac{y^4}{y} = \frac{y^4}{y^1} = y^{4-1} = y^3$$

ANSWER: $-2y^3$

EXAMPLE 2 Divide: $\dfrac{-16c^5d^2}{-8c^2d}$

STEP 1 Divide the coefficients.

$$\frac{-16}{-8} = 2$$

STEP 2 Subtract the exponents of the c variables and the d variables.

$$\frac{c^5}{c^2} = c^{5-2} = c^3$$

$$\frac{d^2}{d} = \frac{d^2}{d^1} = d^{2-1} = d^1 = d$$

ANSWER: $2c^3d$

(**Remember:** If no exponent is written, it is understood to be 1. In Example 1, $y = y^1$. In Example 2, $d = d^1$. Also, d^1 is written as d in the answer.)

Divide. The first problem in each row is done as an example.

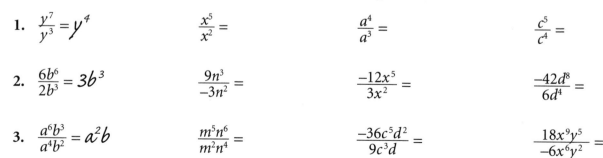

1. $\dfrac{y^7}{y^3} = y^4$ $\dfrac{x^5}{x^2} =$ $\dfrac{a^4}{a^3} =$ $\dfrac{c^5}{c^4} =$

2. $\dfrac{6b^6}{2b^3} = 3b^3$ $\dfrac{9n^3}{-3n^2} =$ $\dfrac{-12x^5}{3x^2} =$ $\dfrac{-42d^8}{6d^4} =$

3. $\dfrac{a^6b^3}{a^4b^2} = a^2b$ $\dfrac{m^5n^6}{m^2n^4} =$ $\dfrac{-36c^5d^2}{9c^3d} =$ $\dfrac{18x^9y^5}{-6x^6y^2} =$

Dividing a Binomial or Trinomial by a Monomial

To divide a binomial or trinomial by a monomial, divide each term and combine the separate quotients. Remove parentheses by dividing each term within parentheses by the monomial.

EXAMPLE 1 Divide: $\dfrac{(4y^3 + 6y)}{2y}$

STEP 1 Divide each term of the binomial by $2y$. Keep the $+$ sign that is between the terms in the binomial.

$$\frac{(4y^3 + 6y)}{2y} = \frac{4y^3}{2y} + \frac{6y}{2y}$$

STEP 2 Divide each pair of terms.

$$\frac{4y^3}{2y} + \frac{6y}{2y} = 2y^2 + 3$$

ANSWER: $2y^2 + 3$

EXAMPLE 2 Divide: $\dfrac{(9a^2b^4 - 3ab^3 + 6b^2)}{3b^2}$.

STEP 1 Divide each term of the trinomial by $3b^2$. Keep the signs that are between the terms in the trinomial.

$$\frac{(9a^2b^4 - 3ab^3 + 6b^2)}{3b^2} =$$

STEP 2 Divide each pair of terms.

$$\frac{9a^2b^4}{3b^2} - \frac{3ab^3}{3b^2} + \frac{6b^2}{3b^2} =$$

$$3a^2b^2 - ab + 2$$

ANSWER: $3a^2b^2 - ab + 2$

Divide. The first problem in each group is done as an example.

1. $\dfrac{(6x^4 + 4x^3)}{2x} = 3x^3 + 2x^2$

$\dfrac{(12n^5 + 6n^4)}{3n^2} =$

$\dfrac{(15y^7 - 9y^5)}{3y^3} =$

2. $\dfrac{(-18n^5 + 6n^4)}{3n} = -6n^4 + 2n^3$

$\dfrac{(-20x^7 + 10x^5)}{-5x^3} =$

$\dfrac{(-14y^5 - 7y^3)}{7y^2} =$

3. $\dfrac{(4a^2b^4 - 2a^2b)}{2a} = 2ab^4 - 2ab$

$\dfrac{(9m^3n + 6m^2n^2)}{3mn} =$

$\dfrac{(16c^2d^2 - 12c^2d)}{4cd} =$

4. $\dfrac{(12a^3b^2 - 6a^2b^3 + 8a^2)}{2a^2} = 6ab^2 - 3b^3 + 4$

$\dfrac{(15m^3n^3 + 10m^2n - 5mn^2)}{5m} =$

$\dfrac{(21c^3d^2 - 7c^2d^2 + 14cd^2)}{7cd} =$

Polynomials Review

Solve each problem below. When you finish, check your answers at the back of the book. Then correct any errors.

Name each polynomial as a monomial, binomial, or trinomial.

1. $3x + 7$ _____

2. $x^2 + 2x + 1$ _____

3. $5y^3$ _____

4. $2z^2 - 5z$ _____

Circle the pair of like terms in each group.

5. $x^2, 2x, -x^3, -5x$

6. $n^3, 3n, 2n^2, -2n^3$

7. $cd^2, -cd^2, c^2d, -c^2d^2$

8. $3ab^2, -2a^2b, a^2b^2, 3a^2b$

Add.

9. $7r + 4r =$

10. $3y^2 + (-y^2) =$

11. $(9x + -4) + x =$

12. $(3n^2 + 5n) + (n^2 - 7n) =$

Write the perimeter of each figure as a polynomial.

13.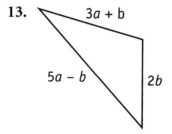

14.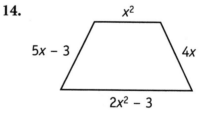

Subtract.

15. $9s - 5s =$

16. $12d - (-5d) =$

17. $(7a + 2b) - 4a =$

18. $(12x - 3y) - (6x + 7y) =$

Write each unlabeled length as a polynomial.

19.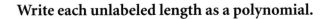

3m + 2n   ?

9m − n

20.

$x^2 + 4x$ $2x^2 + x$

$P = 5x^2 + 6x + 8$

$x^2 + 3$?

Multiply.

21. $5c^4 \cdot 4c^4 =$

22. $-3a^2b(-2ab^3) =$

23. $2x(4x^3 + 3) =$

24. $2m(4m^2 + mn^2 - 3n^3) =$

25. $(3a + 2)(2a - 5) =$

26. $(x + 4)(x - 5) =$

Write the area of each rectangle as a polynomial.

27.
5x

7x + 6

28.
3ab

$a^2 - 2b$

Divide.

29. $\dfrac{n^5}{n^3} =$

30. $\dfrac{12a^7b^5}{(-3a^3b^2)} =$

31. $\dfrac{(15c^4 + 6c^3)}{3c^2} =$

32. $\dfrac{(14m^2n^2 - 8m^2n + 6mn^2)}{2mn} =$

FACTORING

Factors

A **factor** of a number is a whole number that divides evenly into that number. The number 6 has four factors: 1, 2, 3, and 6. Every whole number greater than 1 has at least two factors: itself and the number 1.

Most whole numbers have three or more factors.
- The number 2 has only two factors: 1 and 2.
- The number 8 has four factors: 1, 2, 4, and 8.
- The number 16 has five factors: 1, 2, 4, 8, and 16.

Every whole number greater than 1 can be written as a product of two or more of its factors.

Number:	2	6		12			
Product of Factors:	2×1	6×1	3×2	12×1	6×2	4×3	$3 \times 2 \times 2$

Write all the factors of each number. The first problem in each row is done as an example.

1. 3   9   14
    1, 3

2. 24   36   64
    1, 2, 3, 4, 6, 8, 12, 24

Write each missing factor.

3. $5 = \underline{\qquad} \times 1$ $4 = \underline{\qquad} \times 2$ $8 = \underline{\qquad} \times 2$ $15 = \underline{\qquad} \times 3$

4. $20 = \underline{\qquad} \times 4$ $56 = \underline{\qquad} \times 8$ $60 = \underline{\qquad} \times 4$ $77 = \underline{\qquad} \times 11$

Each number can be written as a product of three factors, none of which is the number 1. Write the two missing factors.

5. $8 = \underline{\qquad} \times \underline{\qquad} \times 2$ $18 = \underline{\qquad} \times \underline{\qquad} \times 2$ $20 = \underline{\qquad} \times \underline{\qquad} \times 2$

6. $30 = \underline{\qquad} \times \underline{\qquad} \times 5$ $42 = \underline{\qquad} \times \underline{\qquad} \times 3$ $50 = \underline{\qquad} \times \underline{\qquad} \times 5$

Prime Numbers and Prime Factors

A **prime number** is any whole number greater than 1 that has only two factors: itself and the number 1. Examples of prime numbers are 2, 3, 5, and 7. A **composite number** is any number greater than 1 that has more than two factors. Examples of composite numbers are 4, 6, 8, and 9.

Prime Number	Factors
2	1, 2
3	1, 3
5	1, 5
7	1, 7

Composite Numbers	Factors
4	1, 2, 4
6	1, 2, 3, 6
8	1, 2, 4, 8
9	1, 3, 9

The number 1 is neither a prime number nor a composite number.

Every composite number can be written as a product of **prime factors** (factors that are prime numbers). Writing a number as a product of prime numbers is called **prime-factorization** form.

Number	Find two factors.	Write each factor as a product of prime factors. (prime-factorization form)	Write each product using exponents.
8	2 and 4	$2 \times 2 \times 2$	2^3
12	2 and 6	$2 \times 2 \times 3$	$2^2 \cdot 3$
36	4 and 9	$2 \times 2 \times 3 \times 3$	$2^2 \cdot 3^2$
70	10 and 7	$2 \times 5 \times 7$	$2 \cdot 5 \cdot 7$
625	25 and 25	$5 \times 5 \times 5 \times 5$	5^4

Write all factors of each number. Circle *prime* or *composite* for each number.

1. 11
   prime or composite

2. 10
   prime or composite

3. 19
   prime or composite

4. 21
   prime or composite

5. 22
   prime or composite

6. 29
   prime or composite

Write each number in prime-factorization form or using exponents.

7. 16

8. 20

9. 32

10. 56

11. 75

12. 96

Finding the Square Root of an Algebraic Term

Sometimes you may be asked to find the square root of an algebraic term. When the term is a perfect square, the square root is easy to find.

- A perfect square can be written as the product of two identical factors. For example, $4x^2$ is a perfect square, being the product of $2x$ times $2x$. The factor $2x$ is the square root of $4x^2$.

An algebraic term is a perfect square if the coefficient is a perfect square and if each exponent is an even number.

- The coefficient of the square root is the square root of the coefficient of the term.
- Each exponent is half the value of the same exponent in the term.

> A factor of an algebraic term or expression may itself be a term or an expression.

To simplify the discussion on these next two pages, only positive square roots will be discussed.

EXAMPLE 1 Find $\sqrt{x^2}$.

ANSWER: x $(x \cdot x = x^2)$

(**Note:** $x = x^1$ and 1 is half of 2.)

EXAMPLE 3 Find $\sqrt{a^6}$.

ANSWER: a^3 $(a^3 \cdot a^3 = a^6)$

(**Note:** 3 is half of 6.)

EXAMPLE 2 Find $\sqrt{25x^6 y^4}$.

ANSWER: $5x^3y^2$ $(5x^3y^2 \cdot 5x^3y^2 = 25x^6y^4)$

(**Note:** $5 = \sqrt{25}$; 3 is half of 6; and 2 is half of 4.)

EXAMPLE 4 Find $\sqrt{16n^4}$.

ANSWER: $4n^2$ $(4n^2 \cdot 4n^2 = 16n^4)$

(**Note:** $4 = \sqrt{16}$, and 2 is half of 4.)

Find the positive square root of each algebraic term.

1. $\sqrt{y^2} =$ $\sqrt{n^4} =$ $\sqrt{y^8} =$ $\sqrt{r^6} =$

2. $\sqrt{25x^4} =$ $\sqrt{16y^8} =$ $\sqrt{36n^6} =$ $\sqrt{49a^{10}} =$

3. $\sqrt{4x^4y^6} =$ $\sqrt{25a^8b^2} =$ $\sqrt{81c^6d^4} =$ $\sqrt{64x^6y^2} =$

Simplifying a Square Root

Sometimes it is possible only to simplify the square root of a number or an algebraic term. Part of the answer remains inside the square root sign. Factoring is used to simplify the square root as much as possible. As a first step, learn this rule for square roots:

> **Rule:** The square root of a product is equal to the product of square roots.

To see that this rule is true, consider $\sqrt{100}$ (which you know has the value 10).

STEP 1 Write 100 as 4×25 and take the square root of this product. $\sqrt{100} = \sqrt{4 \times 25}$

STEP 2 Separate the square root into the product of square roots. $\sqrt{4} \times \sqrt{25}$

STEP 3 Find each square root and multiply the results. $2 \times 5 = 10$

Applying the rule gives the correct value 10 for $\sqrt{100}$.

<u>EXAMPLE 1</u> Simplify: $\sqrt{90}$

STEP 1 Write 90 as $9 \cdot 10$.
$$\sqrt{90} = \sqrt{9 \cdot 10}$$

STEP 2 Separate the square root into the product of square roots.
$$\sqrt{9 \cdot 10} = \sqrt{9} \cdot \sqrt{10}$$

STEP 3 Find each square root and multiply the results
$$\sqrt{9} \cdot \sqrt{10} = 3 \cdot \sqrt{10} = 3\sqrt{10}$$

ANSWER: $3\sqrt{10}$

<u>EXAMPLE 2</u> Simplify: $\sqrt{x^4 y}$

STEP 1 Write $x^4 y$ as $x^4 \cdot y$.
$$\sqrt{x^4 y} = \sqrt{x^4 \cdot y}$$

STEP 2 Separate the square root into the product of square roots.
$$\sqrt{x^4 \cdot y} = \sqrt{x^4} \cdot \sqrt{y}$$

STEP 3 Find each square root and multiply the results.
$$\sqrt{x^4} \cdot \sqrt{y} = x^2 \cdot \sqrt{y} = x^2\sqrt{y}$$

ANSWER: $x^2\sqrt{y}$

Simplify each square root. As the first step, find a factor for each number or power that is a perfect square.

1. $\sqrt{32} = \sqrt{16} \cdot \sqrt{2}$ $\sqrt{45} =$ $\sqrt{75} =$
   $= 4\sqrt{2}$

2. $\sqrt{x^2 y} = \sqrt{x^2} \cdot \sqrt{y}$ $\sqrt{x^6 y} =$ $\sqrt{ab^8} =$
   $= x\sqrt{y}$

3. $\sqrt{4xy^2} = \sqrt{4} \cdot \sqrt{y^2} \cdot \sqrt{x}$ $\sqrt{16 yz^6}$ $\sqrt{25r^4 s} =$
   $= 2y\sqrt{x}$

Factoring an Algebraic Expression

An algebraic expression can often be written as a product of factors.
If each term of an expression can be divided evenly by a number or by a variable, that number or variable is a factor of the expression.

Factoring Out a Number

<u>EXAMPLE 1</u> Write $4x + 10$ as a product of factors.

> **STEP 1** Find the greatest number that divides evenly into each term ($4x$ and 10). The number is 2.
>
> **STEP 2** Write 2 on the outside of a set of parentheses. The result of dividing each term by 2 goes inside the parentheses.

> To factor an algebraic expression is to write the expression as a product of factors.

ANSWER: $4x + 10 = 2(2x + 5)$

Factoring Out a Variable

<u>EXAMPLE 2</u> Write $5x^2 + 3x$ as a product of factors.

> **STEP 1** Each term can be divided evenly by x. So, x can be factored out.
>
> **STEP 2** Write x on the outside of a set of parentheses. For the inside terms, divide $5x^2$ by x ($= 5x$) and divide $3x$ by x ($= 3$).

> The expression $5x^2 + 3x$ contains two terms. Because there is no number that divides evenly into both coefficients, no number can be factored out.

ANSWER: $5x^2 + 3x = x(5x + 3)$

Factor a number out of each algebraic expression.

1. $2x + 6 = 2(x + 3)$ $3x + 12 =$ $5y - 20 =$
   (factor out a 2)

2. $3x^2 + 6x + 3 = 3(x^2 + 2x + 1)$ $4x^2 - 8x + 12 =$ $7y^2 - 14y - 35 =$
   (factor out a 3)

Factor a variable out of each algebraic expression.

3. $2x^2 + 5x = x(2x + 5)$ $3x^2 - 7x =$ $2z^2 + 11z =$
   (factor out an x)

4. $y^3 - 2y^2 + 4y = y(y^2 - 2y + 4)$ $z^3 + 6z^2 - 4z =$ $x^3 - 3x^2 + 2x =$
   (factor out a y)

Factoring Out a Number *and* a Variable

EXAMPLE 3   Write $4x^2 - 12x$ as a product of factors.

> **STEP 1**   Find the greatest number that divides evenly into each term. This number is 4.
>
> **STEP 2**   Notice that you can factor an x out of each term. Factor out both 4 and x. To do this, factor out the product $4x$.

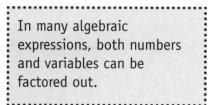

In many algebraic expressions, both numbers and variables can be factored out.

ANSWER: $4x^2 - 12x = 4x(x - 3)$

EXAMPLE 4   Write $3x^3y + 15x^2y$ as a product of factors.

> **STEP 1**   Find the greatest number that divides evenly into each term. This number is 3.
>
> **STEP 2**   Notice that you can also factor the product x^2y out of each term. Factor out $3x^2y$.

A product of variables may be a factor in an algebraic expression.

ANSWER: $3x^3y + 15x^2y = 3x^2y \,(x + 5)$

Factor each algebraic expression as completely as possible.

5. $2w^2 + 8 = 2(w^2 + 4)$ $9x^2 + 3 =$ $12y^2 - 8 =$
    (factor out 2)

6. $5x^2 + 10x = 5x(x + 2)$ $6x^2 + 24x =$ $7z^2 - 21z =$
    (factor out 5x)

7. $12z^2 - 18z = 6z(2z - 3)$ $4y^2 + 6y =$ $10z^2 - 15z =$
    (factor out 6z)

8. $3y^3 + 15\,y^2 = 3y^2(y + 5)$ $5x^3 - 25x^2 =$ $4s^3 + 16s^2$
    (factor out 3y²)

9. $8x^4 + 12x^2 = 4x^2(2x^2 + 3)$ $9y^4 + 6y^2 =$ $15z^4 + 20z^2 =$
    (factor out 4x²)

10. $2ab^2 + 4a^2b = 2ab(b + 2a)$ $6x^2y + 3xy =$ $4r^2s + 6rs^2 =$
    (factor out 2ab)

Factoring and Dividing

Factoring is often used to simplify a quotient of algebraic expressions. If a single term or an expression of 2 or more terms is a factor in both top and bottom expressions, that term or expression divides and cancels out.

EXAMPLE 1 Divide: $\dfrac{6n^2 + 2}{2}$

  STEP 1 Factor 2 out $6n^2 + 2$.

     $6n^2 + 2 = 2(3n^2 + 1)$

  STEP 2 Divide out the 2's.

     $\dfrac{6n^2 + 2}{2} = \dfrac{2(3n^2 + 1)}{\cancel{2}}$

ANSWER: $3n^2 + 1$

EXAMPLE 2 Divide: $\dfrac{3x^2 + 6x}{3x}$

  STEP 1 Factor $3x$ out $3x^2 + 6x$.

     $3x^2 + 6x = 3x(x + 2)$

  STEP 2 Divide out the $3x$'s.

     $\dfrac{3x^2 + 6x}{3x} = \dfrac{\cancel{3x}(x + 2)}{\cancel{3x}}$

ANSWER: $x + 2$

EXAMPLE 3 Divide: $\dfrac{8x^2y^2 - 4xy}{4xy}$

  STEP 1 Factor $4xy$ out $8x^2y^2 - 4xy$.

     $8x^2y^2 - 4xy = 4xy(2xy - 1)$

  STEP 2 Divide out the term $4xy$.

     $\dfrac{8x^2y^2 - 4xy}{4xy} = \dfrac{\cancel{4xy}(2xy - 1)}{\cancel{4xy}}$

ANSWER: $2xy - 1$

EXAMPLE 4 Divide: $\dfrac{2a^2 + 2ab}{a + b}$

  STEP 1 Factor $2a$ out $2a^2 + 2ab$.

     $2a^2 + 2ab = 2a(a + b)$

  STEP 2 Divide out the expression $a + b$.

     $\dfrac{2a^2 + 2ab}{a + b} = \dfrac{2a\cancel{(a + b)}}{\cancel{a + b}}$

ANSWER: $2a$

Simplify each quotient by dividing out the common factor. The first problem in each row is started for you.

1. $\dfrac{4x^2 + 8}{4} = \dfrac{4(\qquad)}{4} =$ $\dfrac{8r^2 + 6}{2} =$

2. $\dfrac{4y^2 + 2y}{2y} = \dfrac{2y(\qquad)}{2y} =$ $\dfrac{2n^2 + 6n}{2n} =$

3. $\dfrac{5x^2y - 10xy}{5xy} = \dfrac{5xy(\qquad)}{5xy} =$ $\dfrac{6ab - 3a^2b^2}{3ab} =$

4. $\dfrac{3x + 3y}{x + y} = \dfrac{3(\underline{\hspace{2cm}})}{x + y} = $

   $\dfrac{4c - 4d}{c - d} = $

5. $\dfrac{x^2 + xy}{x + y} = \dfrac{x(\underline{\hspace{2cm}})}{x + y} = $

   $\dfrac{mn + n^2}{m + n} = $

6. $\dfrac{3r^2 - 3rs}{r - s} = \dfrac{3r(\underline{\hspace{2cm}})}{r - s} = $

   $\dfrac{7uv + 7v^2}{u + v} = $

Write each missing dimension as a simplified algebraic expression.

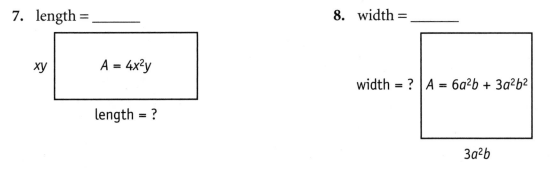

7. length = _____

   xy | $A = 4x^2y$

   length = ?

   $\text{length} = \dfrac{\text{area}}{\text{width}}$

8. width = _____

   width = ? | $A = 6a^2b + 3a^2b^2$

   $3a^2b$

   $\text{width} = \dfrac{\text{area}}{\text{length}}$

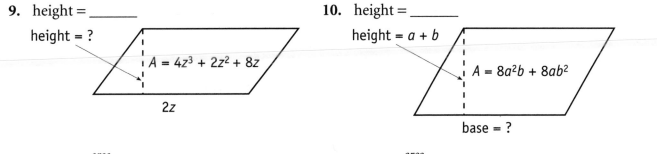

9. height = _____

   height = ?

   $A = 4z^3 + 2z^2 + 8z$

   $2z$

   $\text{height} = \dfrac{\text{area}}{\text{base}}$

10. height = _____

    height = $a + b$

    $A = 8a^2b + 8ab^2$

    base = ?

    $\text{base} = \dfrac{\text{area}}{\text{height}}$

Factoring a Difference of Squares

When you multiply the sum of two variables by the difference of the same two variables, the product is the difference of the squares of the variables.

$$(a + b)(a - b) = a^2 - b^2 \qquad [(a + b)(a - b) = a^2 + ab - ab - b^2 = a^2 - b^2]$$

Knowing this enables you to write the rule for factoring a difference of squares.

Rule for factoring a difference of squares:
A difference of squares factors into two binomials: the sum of the variables *times* the difference of the variables.

This rule applies to numbers as well as variables.

EXAMPLE 1 Factor $x^2 - 4$.

> **STEP 1** Find the square roots of x^2 and 4.
> ($\sqrt{x^2} = x$) and 4 ($\sqrt{4} = 2$)
>
> **STEP 2** Write the difference of squares as a product of binomial factors.
> $x^2 - 4 = (x + 2)(x - 2)$

ANSWER: $x^2 - 4 = (x + 2)(x - 2)$

EXAMPLE 2 Factor $16 - r^2$.

> **STEP 1** Find the square roots of 16 and r^2
> ($\sqrt{16} = 4$) and r^2 ($\sqrt{r^2} = r$)
>
> **STEP 2** Write the difference of squares as a product of binomial factors.
> $16 - r^2 = (4 + r)(4 - r)$

ANSWER: $16 - r^2 = (4 + r)(4 - r)$

Factor each difference of squares.

1. $25 - s^2 =$ $\qquad\qquad\qquad\qquad$ $36 - t^2 =$

2. $64 - n^2 =$ $\qquad\qquad\qquad\qquad$ $16 - x^2 =$

3. $r^2 - 49 =$ $\qquad\qquad\qquad\qquad$ $y^2 - 81 =$

Completely factor each expression. The first problem in each row is done as an example.

4. $x^3 - 25x = x(x^2 - 25) = x(x + 5)(x - 5)$

   $y^3 - 36y =$

   $n^3 - 16n =$

   > For problems 4 and 5, a term is factored out and then the difference of squares is factored.

5. $4s^3 - 36s = 4s(s^2 - 9) = 4s(s + 3)(s - 3)$

   $8x^3 - 32x =$

   $2y^3 - 32y =$

Simplify each quotient by factoring. The first problem is done as an example.

6. $\dfrac{d^2 - 81}{d - 9} = \dfrac{(d + 9)(d - 9)}{d - 9} = d + 9$ $\dfrac{t^2 - 64}{t + 8} =$

7. $\dfrac{25 - r^2}{5 + r} =$ $\dfrac{4n^3 - 36n}{n - 3} =$

Write each missing dimension as a simplified algebraic expression. As a first step, completely factor the expression for each area.

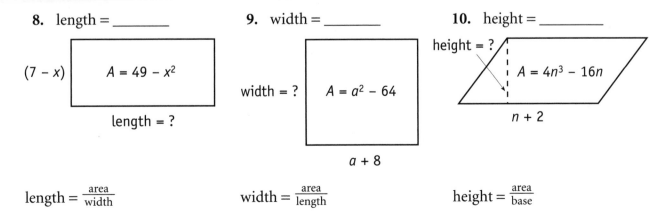

8. length = _____

   $(7 - x)$ | $A = 49 - x^2$

   length = ?

   $\text{length} = \dfrac{\text{area}}{\text{width}}$

9. width = _____

   width = ? | $A = a^2 - 64$

   $a + 8$

   $\text{width} = \dfrac{\text{area}}{\text{length}}$

10. height = _____

    height = ? | $A = 4n^3 - 16n$

    $n + 2$

    $\text{height} = \dfrac{\text{area}}{\text{base}}$

Factoring Review

Solve the problems below. When you finish, check your answers at the back of the book. Then correct any errors.

Write all the factors of each number.

1. 9

2. 7

3. 24

Each number can be written as a product of three factors, none of which is the number 1. Write the two missing factors.

4. $24 = \underline{\quad} \times \underline{\quad} \times 3$

5. $30 = \underline{\quad} \times \underline{\quad} \times 5$

6. $42 = \underline{\quad} \times \underline{\quad} \times 2$

Write each number in prime-factorization form.

7. 14

8. 16

9. 50

Find the positive square root of each term.

10. $\sqrt{16x^4} =$

11. $\sqrt{36a^2b^2}$

12. $\sqrt{25c^4d^6}$

Simplify each square root.

13. $\sqrt{18} =$

14. $\sqrt{25x^2y}$

15. $\sqrt{49cd^6}$

Factor a number out of each expression.

16. $4x + 4 =$

17. $3y - 15 =$

18. $6y + 24 =$

Factor a variable out of each expression.

19. $3x^2 + 7x =$

20. $2n^3 - 9n =$

21. $5y^4 + 2y^2 =$

Factor each expression as completely as possible.

22. $3x^2 + 12x =$

23. $5n^3 - 15n^2 =$

24. $8x^2y + 2xy =$

Simplify each quotient by dividing out the common factor.

25. $\dfrac{4x + 4}{4} =$

26. $\dfrac{3n^2 + 6}{3} =$

27. $\dfrac{6m - 3m^2n}{3m} =$

28. $\dfrac{4x + 4y}{x + y} =$

29. $\dfrac{xy + y^2}{x + y} =$

30. $\dfrac{3ab - 3b^2}{a - b} =$

Factor each difference of squares.

31. $36 - n^2 =$

32. $x^2 - 25 =$

Simplify each quotient by factoring.

33. $\dfrac{x^2 - 49}{x + 7} =$

34. $\dfrac{3n^2 - 27}{n - 3} =$

Write each missing dimension as a simplified algebraic expression.

35. length = _____

36. width = _____

$4ab$ | $A = 12a^2b^2 + 4ab^2$

length = ?

width = ? | $A = x^2 - 36$

$x + 6$

Algebra Posttest A

This posttest gives you a chance to check the skills you have learned in *Algebra.* Take your time and work each problem carefully. When you finish, check your answers and review any topics on which you need more work.

Write the addition shown by the number arrows.

1. _____

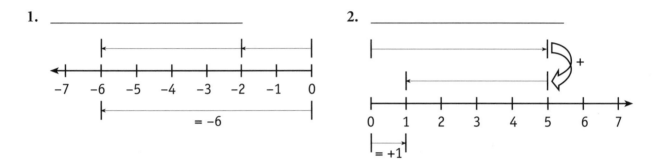

2. _____

Subtract.

3. $8 - (-7) =$

4. $-4 - (-7) =$

Multiply or divide as indicated.

5. $(-8)(4) =$

6. $\dfrac{-5}{-9} =$

Find the value of each power.

7. $3^4 =$

8. $4^{-3} =$

Simplify each product or quotient.

9. $(7^3)(7^2)(7) =$

10. $\dfrac{10^8}{10^4} =$

Find each square root below.

11. $\sqrt{81} =$

12. $\sqrt{\dfrac{36}{25}} =$

13. Write the expression "$\frac{1}{4}$ times the quantity x minus 6" as an algebraic expression.

Find the value of each expression.

14. $5n^2 - 3n$ when $n = 2$

15. $ab\left(\dfrac{a+4}{b-4}\right)$ when $a = -2, b = 6$

16. Use the formula $°C = \frac{5}{9}(°F - 32)$ to find the Celsius temperature ($°C$) when the Fahrenheit temperature ($°F$) is 77°F.

Circle the equation that is *not* represented in the drawing.

17.

a. $n - 4 = 10$
b. $10 - n = 4$
c. $n + 4 = 10$

For problems 18–22, solve each equation. Show all steps.

18. **a.** $n + 12 = 6$ **b.** $x - 7 = 30$

19. **a.** $4p = 36$ **b.** $\frac{r}{6} = 11$

20. **a.** $4x - 7 = 29$ **b.** $\frac{a}{3} + 4 = 9$

21. **a.** $6n + 2n = 72$ **b.** $5x - 8 = x + 16$

22. **a.** $3(s - 2) = 33$ **b.** $5(y - 2) = 2(y + 13)$

23. A field is in the shape of a rectangle. The ratio of the length of the field to its width is 3 to 2. If the perimeter of the field is 190 yards, what are the dimensions of the field?

24. In a survey of the 360 high school students, 3 out of every 5 said they ride the school bus at least four times each week. Use a proportion to find the number of students who do *not* ride the school bus at least four times each week.

25. Find the two correct values of x in the equation $4x^2 = y$ when $y = 144$.

26. Graph the line containing the points $(-3, 0)$, $(0, 1)$, and $(3, 2)$.

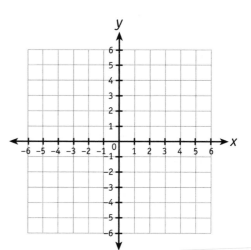

27. What is the slope of line R?

$$\text{slope} = \frac{\text{change in } y}{\text{change in } x}$$

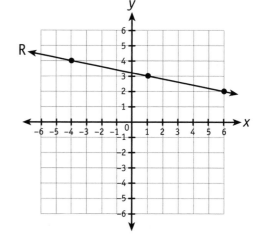

28. Graph the equation $y = 3x - 4$. As a first step, complete the Table of Values.

Table of Values

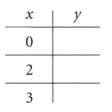

x	y
0	
2	
3	

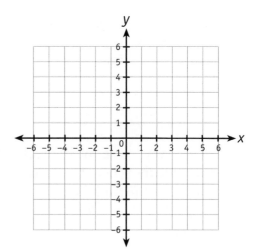

Write the inequality graphed on the number line.

29.

Values of n

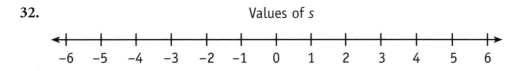

Solve the following inequalities.

30. $x - 8 \geq -2$

31. $4r + 7 < 35$

Graph the inequality $-4 \leq s < 5$.

32.

Values of s

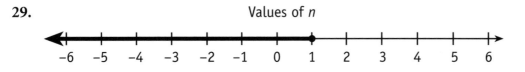

33. Circle the pair of like terms.

$2y^3 \qquad 2x^3 \qquad 3y^2 \qquad -3x^2 \qquad -3x^3$

Add or subtract as indicated.

34. $(-4n^2 + 7) + (5n^2 - 9) =$

35. $(5x^2 - 6x) - (x^2 + 3x) =$

Multiply or divide as indicated.

36. $4d(2d^2 - d + 1) =$

37. $\dfrac{(21n^4 + 15n^2)}{3n}$

Simplify each square root.

38. $\sqrt{72} =$

39. $\sqrt{49\,a^6b} =$

Factor each expression as completely as possible.

40. $9x^2 + 12 =$

41. $20n^3 + 15n^2 =$

Simplify each quotient by dividing out the common factor.

42. $\dfrac{8x^2 + 4}{4} =$

43. $\dfrac{9a - 9b}{a - b} =$

Algebra Posttest A Prescriptions

Circle the number of any problem that you miss. A passing score is 37 correct answers. If you passed the test, go on to Using Number Power. If you did not pass the test, review the chapters in this book or refer to these practice pages in other materials from Contemporary Books.

PROBLEM NUMBERS	SKILL AREA	PRACTICE PAGES
1, 2, 3, 4, 5, 6	signed numbers	10–21
	Real Numbers: Algebra Basics	30–50
7, 8, 9, 10, 11, 12	powers and roots	24–35
	Math Exercises: Pre-Algebra	4–7
13, 14, 15, 16	algebraic expressions	38–49
	Real Numbers: Algebra Basics	1–4, 60–63
	Math Exercises: Pre-Algebra	8–21
17, 18, 19	one-step equations	52–63
	Real Numbers: Algebra Basics	5–20
20, 21, 22	multistep equations	68–81
	Real Numbers: Algebra Basics	22–28
	Math Exercises: Algebra	4–14
23, 24, 25	ratios, proportions, and quadratics	86–97
	Math Exercises: Pre-Algebra	24–27
26, 27, 28	graphing equations	100–113
	Math Exercises: Algebra	20–27
29, 30, 31, 32	inequalities	116–125
33, 34, 35, 36, 37	polynomials	128–139
38, 39, 40, 41, 42, 43	factoring	142–151

For further algebra practice:

Pre-GED/Basic Skills Interactive (software)
Mathematics Units 4 and 6

GED Interactive (software)
Mathematics Unit 6

Algebra Posttest B

This posttest gives you a chance to check your algebra skills in a multiple-choice format as used in the GED test and other standardized tests. Take your time and work each problem carefully before you choose your answer. When you finish, check your answers with the answer key at the back of the book.

1. Which addition is shown by the number arrows?

   a. $0 + 4$
   b. $4 + 1$
   c. $3 + (-4)$
   d. $4 + (-1)$
   e. $4 + (-3)$

   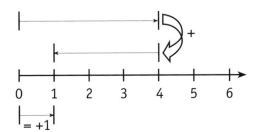

2. At 7:00 P.M., the temperature in Anchorage, Alaska, was –9°F. By 3:00 A.M., the temperature had dropped another 15°F. What was the temperature at 3:00 A.M.?

   a. –6°F　　　b. –15°F　　　c. –24°F　　　d. +6°F　　　e. +9°F

3. What is the sum of the product $(-3)(-7)$ and the quotient $(-10)/2$?

   a. 26　　　b. 16　　　c. 0　　　d. –16　　　e. –26

4. What is the value of $(-2)^4$?

   a. –16　　　b. –8　　　c. –4　　　d. 8　　　e. 16

5. Which power equals the quotient $\dfrac{6^8}{6^3}$?

   a. 6^5　　　b. 6^{11}　　　c. 6^{24}　　　d. 6^{-5}　　　e. 6^{-11}

6. Find the square root. $\sqrt{\frac{36}{25}} =$

   a. $\frac{12}{5}$ b. $\frac{6}{5}$ c. $\frac{6}{25}$ d. $\frac{25}{6}$ e. $\frac{5}{6}$

7. Which algebraic expression is read "4 times the quantity n minus 7"?

   a. $4n - 7$ b. $4 - 7n$ c. $4(n - 7)$ d. $4(7 - n)$ e. $4(-7n)$

8. What is the value of the algebraic expression below?

   $(a^2 + 1)(b^2 - 1)$ when $a = 2, b = 3$

   a. 28   b. 32   c. 36   d. 40   e. 44

9. According to the formula $°C = \frac{5}{9}(°F - 32)$, what is the Celsius temperature (°C) when the Fahrenheit temperature (°F) is 50°F?

   a. 10°C   b. 18°C   c. 32°C   d. 100°C   e. 212°C

10. Which equation is *not* represented by the drawing?

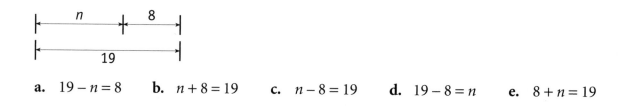

    a. $19 - n = 8$ b. $n + 8 = 19$ c. $n - 8 = 19$ d. $19 - 8 = n$ e. $8 + n = 19$

11. What is the first step in solving the equation $y - 8 = 23$?

 a. Subtract +8 from each side of the equation.
 b. Multiply each side of the equation by +8.
 c. Divide each side of the equation by +8.
 d. Add −8 to each side of the equation.
 e. Add +8 to each side of the equation.

12. What is the first step in solving the equation $3x + 9 = 15$?

 a. Divide 15 by 3.
 b. Add 9 to each side of the equation.
 c. Multiply each side of the equation by 3.
 d. Subtract 9 from each side of the equation.
 e. Subtract $2x$ from each side of the equation.

13. What is the solution to the equation $3x - 8 = 46$?

 a. $x = 18$ b. $x = 16$ c. $x = 14$ d. $x = 12$ e. $x = 10$

14. What is the solution to the equation $2(n + 4) = n + 12$

 a. $n = 0$ b. $n = 2$ c. $n = 4$ d. $n = 6$ e. $n = 8$

15. Kim, Yvette, and Jeff have a house painting business. Kim earns $40 less a month than Yvette. Jeff earns twice as much as Kim. During June, the business earned $2,620. Which equation can be used to find the amount earned (x) during June by Kim?

 a. $(x + \$40) + 2x = \$2,620$
 b. $(x - \$40) + 2x = \$2,620$
 c. $x + (x - \$40) + 2x = \$2,620$
 d. $x + (x + \$40) + 2x = \$2,620$
 e. $2x + (x + \$40) - x = \$2,620$

16. The ratio of two numbers is 5 to 4. The sum of the numbers is 99. What number is the greater of the two numbers?

    a. 33   b. 44   c. 55   d. 66   e. 77

17. For how many different values of x is the equation $3x^2 = y$ true when $y = 75$?

    a. none   b. one   c. two   d. three   e. four

18. Jani makes punch by mixing 3 cups of pineapple juice with 5 cups of orange juice. Which proportion tells how many cups (n) of pineapple juice should Jani mix with 18 cups of orange juice?

    a. $\dfrac{n}{3} = \dfrac{5}{18}$ b. $\dfrac{n}{5} = \dfrac{3}{18}$ c. $\dfrac{n}{18} = \dfrac{5}{8}$ d. $\dfrac{n}{18} = \dfrac{3}{5}$ e. $\dfrac{n}{18} = \dfrac{3}{8}$

For problems 19 and 20, refer to the grid at the right.

19. Which point is the x-intercept of line L?

    a. $(0, 4)$
    b. $(2, 2)$
    c. $(0, 0)$
    d. $(6, -2)$
    e. $(4, 0)$

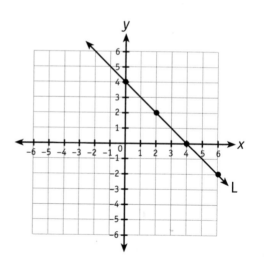

20. What is the slope of line L?

    $$\text{slope} = \frac{\text{change in } y}{\text{change in } x}$$

    a. -2
    b. -1
    c. 0
    d. 1
    e. 2

21. Which point does *not* lie on the line of solutions of the graph of the linear equation $y = 2x - 3$?

   a. $(-1, -5)$ **b.** $(0, -3)$ **c.** $(1, 2)$ **d.** $(2, 1)$ **e.** $(3, 3)$

22. Which number is *not* an allowed value for x in the inequality $2x - 6 \leq 5$?

   a. -6 **b.** -4 **c.** 0 **d.** 4 **e.** 6

23. Which inequality expresses the relationship represented by the line drawing?

   a. $3x + 6 < 35$ **d.** $3x + 6 = 35$
   b. $3x + 6 > 35$ **e.** $3x + 6 \leq 35$
   c. $3x + 6 \geq 35$

24. Which inequality is graphed on the number line?

   a. $-4 \leq x + 4$ **d.** $-4 \leq x \leq 4$
   b. $-4 < x \leq 4$ **e.** $-4 \leq x < 4$
   c. $-4 < x < 4$

25. Subtract: $(6n^2 - n) - (3n^2 + 2n)$

   a. $3n^2 - 3n$ **d.** $9n^2 + n$
   b. $3n^2 + 3n$ **e.** $9n^2 - 3n$
   c. $3n^2 + n$

26. Divide: $\dfrac{(26x^5 + 16x^3)}{2x^3}$

   a. $13x^2 + 8x^3$ **d.** $13x^2 + 16x$
   b. $13x^2 + 16x^3$ **e.** $13x^2 + 8$
   c. $13x^2 + 8x$

27. What is the volume of the rectangular solid?

a. $x^2 - 2x$
b. $x^2 + 2x$
c. $x^4 - 4x^3$
d. $x^4 + 4x^3$
e. $x^4 - 4x$

28. Simplify: $\sqrt{49\,x^6 y}$

a. $7x^3\sqrt{y}$
b. $7x^3\sqrt{3y}$
c. $7x^2\sqrt{y}$
d. $7x\sqrt{y}$
e. $7x\sqrt{3y}$

29. Factor the expression $16n^4 - 24n^2$ as completely as possible.

a. $n^2(16n^2 - 24)$
b. $n^2(16n^2 - 48)$
c. $4n^2(4n^2 - 6)$
d. $8n^2(2n^2 - 3)$
e. $8n^2(2n^2 - 3n)$

30. Simplify the quotient $\dfrac{r^2 - 9}{r + 3}$.

a. $r^2 + 3$
b. $r^2 - 3$
c. $3r^2$
d. $r + 3$
e. $r - 3$

Algebra Posttest B Chart

If you missed more than one problem in any group below, review the practice pages for those problems. Then redo the problems you got wrong before going on to Using Number Power. If you had a passing score, redo any problems you missed and go on to Using Number Power on page 167.

PROBLEM NUMBERS	SKILL AREA	PRACTICE PAGES
1, 2, 3	signed numbers	10–21
4, 5, 6	powers and roots	24–35
7, 8, 9	algebraic expressions	38–49
10, 11	one-step equations	52–63
12, 13, 14, 15	multistep equations	68–81
16, 17, 18	ratios, proportions, and quadratics	86–97
19, 20, 21	graphing equations	100–113
22, 23, 24	inequalities	116–125
25, 26, 27	polynomials	128–139
28, 29, 30	factoring	142–151

Using
Number
Power

Wind Chill

When the wind blows, your body feels colder than the actual temperature. For example, a wind speed of 25 miles per hour causes a 10°F temperature to feel as cold as a temperature of –29°F in which no wind is blowing. **Wind chill** is the name given to the temperature that your body actually feels when the wind is blowing.

The table below shows thermometer temperatures and the wind chill temperature for various wind speeds. The example given above is highlighted in the table.

		Thermometer Temperature (°F)										
		–25°	–20°	–15°	–10°	–5°	0°	5°	10°	15°	20°	25°
		Wind Chill Temperature (°F)										
	5	–31°	–26°	–21°	–15°	–10°	–5°	0°	6°	11°	16°	22°
	10	–52°	–46°	–40°	–34°	–27°	–22°	–15°	–9°	–3°	3°	10°
Wind	15	–65°	–58°	–51°	–45°	–38°	–31°	–25°	–18°	–11°	–5°	2°
Speed	20	–74°	–67°	–60°	–53°	–46°	–39°	–31°	–24°	–17°	–10°	–3°
(mph)	25	–81°	–74°	–66°	–59°	–51°	–44°	–36°	–29°	–22°	–15°	–7°
	30	–86°	–79°	–71°	–64°	–56°	–49°	–41°	–33°	–25°	–18°	–10°
	35	–89°	–82°	–74°	–67°	–58°	–52°	–43°	–35°	–27°	–20°	–12°
	40	–92°	–84°	–76°	–69°	–60°	–53°	–45°	–37°	–29°	–21°	–13°

EXAMPLE When the wind speed is 35 mph, how many degrees below the thermometer temperature of 25°F is the wind chill temperature?

> **STEP 1** On the table, read down the column beneath the temperature 25°F. Read across from the row of wind speed 35 mph until it meets the 25° temperature column. This temperature is –12°F.
>
> **STEP 2** Subtract –12 from 25.
>
> $$25 - (-12) = 25 + 12 = \mathbf{37}$$

ANSWER: The wind chill temperature is **37°** below the thermometer temperature.

Use the table on page 168 to find the wind chill temperature for each thermometer temperature.

1. thermometer temperature = −20°F
   wind speed = 30 mph

   wind chill temperature =

2. thermometer temperature = −5°F
   wind speed = 10 mph

   wind chill temperature =

3. thermometer temperature = 0°F
   wind speed = 40 mph

   wind chill temperature =

4. thermometer temperature = 20°F
   wind speed = 15 mph

   wind chill temperature =

Use the table on page 168 to find the difference between the thermometer temperature and the wind chill temperature (thermometer temperature minus wind chill temperature). As a first step, find the wind chill temperature.

5. thermometer temperature = −15°F
   wind speed = 20 mph

   difference between thermometer
   and wind chill temperatures =

6. thermometer temperature = 15°F
   wind speed = 30 mph

   difference between thermometer
   and wind chill temperatures =

Use the table on page 168 to solve each problem.

7. A weather reporter in Chicago says that the wind chill temperature is −17°F. At the same time, Kristen reads an outdoor thermometer which gives a temperature of 15°F. At this moment, what is the wind speed?

8. While standing in a 10 mph wind, Sean hears a report that the wind chill temperature is −27°F. At this moment, what is the actual thermometer temperature?

Distance Formula

The distance formula, $d = rt$, tells how to find the distance traveled (d) when you know the rate (r) and the time of travel (t).

EXAMPLE 1   How far can Joan drive in 7 hours if she averages 50 miles per hour?

> **STEP 1**   Identify r and t.
>
> $r = 50$ miles per hour and $t = 7$ hours
>
> **STEP 2**   Substitute the r and t values into the distance formula.
>
> $d = rt = 50 \times 7 = $ **350 miles**

ANSWER: 350 miles

The distance formula can be written in two other ways: as a rate formula and as a time formula.

The rate formula, $r = \frac{d}{t}$, is used to find the rate (speed) when the distance and time are known.

EXAMPLE 2   On part of her trip, Joan drove 405 miles in 9 hours. What was her average speed?

> **STEP 1**   Identify d and t.
>
> $d = 405$ miles and $t = 9$ hours
>
> **STEP 2**   Substitute the d and t values into the rate formula and solve.
>
> $r = \frac{d}{t} = \frac{405}{9} = $ **45 miles per hour**

ANSWER: 45 miles per hour

The time formula, $t = \frac{d}{r}$, is used to find the time when the distance and rate (speed) are known.

EXAMPLE 3   Joan has 280 miles left to drive. If she averages 40 miles per hour, how long will it take her to drive this distance?

> **STEP 1**   Identify d and r.
>
> $d = 280$ miles and $r = 40$ miles per hour
>
> **STEP 2**   Substitute the d and r values into the time formula.
>
> $t = \frac{d}{r} = \frac{280}{40} = $ **7 hours**

ANSWER: 7 hours

For each problem, decide whether you are finding the distance (_d_), rate (_r_), or time (_t_). Then use the appropriate formula to solve.

| distance formula: $d = rt$ | rate formula: $r = \frac{d}{t}$ | time formula: $t = \frac{d}{r}$ |

1. What distance can Bert drive in 12 hours if he averages 50 miles per hour?

2. On the second day of his trip, Bert drove 440 miles in 8 hours. Find his average speed during the second day.

3. If Bert can average 50 miles per hour on the third day, how long will it take him to drive 350 miles?

4. Jennifer drove from Chicago to New York. On the first day of her trip, she averaged 48 miles per hour for 6 hours. How far did she drive the first day?

5. On the second day of her trip, Jennifer drove 364 miles in 7 hours. Find her average speed during the second day.

6. On the third day, Jennifer averaged 54 miles per hour as she drove 324 miles. How many hours did Jennifer drive on the third day?

Simple Interest Formula

Interest is money that is earned (or paid) for the use of someone else's money.

- If you deposit money in a savings account, interest is the money the bank pays you for using your money.
- If you borrow money, interest is the money you pay for using the lender's money.

Simple interest is interest on a principal amount—money deposited or borrowed. To compute simple interest, use the **simple interest formula.** The formula is $i = prt$, where

i = simple interest, written in dollars
p = principal, money deposited or borrowed, written in dollars
r = percent rate, written as a fraction or decimal
t = time, written in years

EXAMPLE 1   What is the simple interest earned on $300 deposited for 2 years in a savings account which pays 5% simple interest?

STEP 1   Identify p, r, and t.

$$p = 300 \quad r = 5\% = \frac{5}{100} \quad t = 2$$

The rate, 5%, is expressed as the fraction $\frac{5}{100}$ to simplify multiplication in Step 2.

STEP 2   Substitute and multiply. Cancel if you can.

$$i = prt = \overset{3}{\cancel{300}} \times \frac{5}{\underset{1}{\cancel{100}}} \times 2 = \textbf{\$30}$$

ANSWER: $30

EXAMPLE 2   At the end of 3 years, what is the total amount owed on a $429 loan borrowed at 9.5% simple interest per year?

STEP 1   Identify p, r, and t.
$$p = 429 \quad r = 9.5\% = 0.095 \quad t = 3$$

Sometimes it is easier to multiply with decimals than with fractions, especially when there would be no cancellation with fractions.

STEP 2   Substitute and multiply.
$$pr = 429 \times 0.095 = 40.755$$
$$prt = 40.755 \times 3 = \$122.265 \approx \$122.27 \text{ (nearest cent)}$$

STEP 3   principal + interest = total amount owed
$$\$429.00 + 122.27 = \textbf{\$551.27}$$

ANSWER: $551.27

For each problem, decide whether you are finding the interest or the new principal (the original principal plus the interest). Then use the interest formula to solve.

1. What is the interest earned on $200 deposited for 3 years in a savings account that pays 5% simple interest?

2. Jane deposited $650 at a $5\frac{1}{2}$% simple interest rate. What will be the total amount in her account after 2 years?

3. At a 10% interest rate, how much simple interest will Bill pay on a loan of $475 borrowed for 2 years?

4. What is the total amount owed on a loan if the principal is $1,000, the simple interest rate is 11%, and the time is 3 years?

5. How much simple interest will John pay for a loan of $575 at 12.5% if he repays the bank at the end of 1 year?

6. What is the total amount in a savings account if the amount deposited was $375, the simple interest rate was 5.25%, and the time was 4 years?

Simple Interest Formula: Parts of a Year

Interest is paid on the basis of a yearly percent rate. However, not all deposits or loans are made for whole years. To use the simple interest formula for a time period that is not a whole year, write the time as a fraction of a year.

EXAMPLE 1 How much simple interest is earned on a $600 deposit at an interest rate of 7% for 8 months?

    STEP 1 Identify p, r, and t.

$$p = 600 \quad r = \frac{7}{100} \quad t = \frac{8}{12} = \frac{2}{3}$$

    (**Note:** 8 months $= \frac{8}{12}$ year $= \frac{2}{3}$ year)

    STEP 2 Substitute and multiply. Cancel if possible.

$$i = prt = \overset{6}{\cancel{600}}\left(\frac{7}{\underset{1}{\cancel{100}}}\right)\left(\frac{2}{3}\right) = 6(7)\left(\frac{2}{3}\right) = \textbf{\$28}$$

ANSWER: $28

EXAMPLE 2 What is the interest paid on a loan of $800 borrowed for 2 years 3 months at 9% simple interest?

    STEP 1 Identify p, r, and t.

$$p = 800 \quad r = \frac{9}{100} \quad t = 2\frac{3}{12} = 2\frac{1}{4} = \frac{9}{4}$$

    (**Note:** Time is written at $\frac{9}{4}$ to simplify multiplication in Step 2.)

    STEP 2 Substitute and multiply. Cancel if possible.

$$i = prt = \overset{8}{\cancel{800}}\left(\frac{9}{\underset{1}{\cancel{100}}}\right)\left(\frac{9}{4}\right) = 8(9)\left(\frac{9}{4}\right) = \textbf{\$162}$$

ANSWER: $162

EXAMPLE 3 What is the interest earned on $200 deposited for 16 months in a savings account paying 6% simple interest?

    STEP 1 Identify p, r, and t.

$$p = 200 \quad r = \frac{6}{100} \quad t = \frac{16}{12} = \frac{4}{3}$$

    STEP 2 Substitute and multiply. Cancel if possible.

$$i = prt = \overset{2}{\cancel{200}}\left(\frac{6}{\underset{1}{\cancel{100}}}\right)\left(\frac{4}{3}\right) = 2(6)\left(\frac{4}{3}\right) = \textbf{\$16}$$

ANSWER: $16

For each problem, decide whether you are finding the interest or the new principal (original principal plus the interest). Then use the simple interest formula to solve.

1. What is the interest earned on $900 deposited for 1 year 6 months in a savings account that pays 5% simple interest?

2. How much money will Tracy repay to the bank for a loan of $1,000 that she borrows for 2 years 8 months at a 12% simple interest rate?

3. Eva deposited $850 in a savings account. How much interest will her money earn after 20 months if the simple interest rate is 6%?

4. How much money will Lam repay to the bank if he borrows $1,000 for 9 months at a simple interest rate of 16%?

5. Top Dollar Visa is charging Keisha 18% simple interest on her $750 unpaid bill. How much interest will Keisha pay if she pays the entire bill after 10 months?

6. Anne plans to take out a $7,000 loan at Evergreen Bank to buy a used car. The bank agrees to let Anne pay the principal plus simple interest at the end of 18 months. What total amount will Anne owe?

Evergreen Bank Simple-Interest Loans	
New car	9.9%
Used car	10.5%
Boat	12.8%

Work Problems

A common problem (often called a **work problem**) involves finding the rate at which two or more people working together can do a job. If you know the rate at which each person works alone, you can find the rate at which they work together.

EXAMPLE   Juan can do a yard job alone in 2 hours. Miah can do the same job alone in 3 hours. How long will it take both of them working together to finish the job?

STEP 1   Express the fraction of the job each can do in 1 hour.

$\frac{1}{2}$ = fraction of the job Juan can do in 1 hour

(If Juan can do the whole job in 2 hours, he can do $\frac{1}{2}$ of the job in 1 hour.)

$\frac{1}{3}$ = fraction of the job Miah can do in 1 hour

STEP 2   Express the fraction of the job each can do in x hours, where x equals the time it takes to complete the job together.

$\frac{1}{2}x$ = fraction of the job Juan can do in x hours

$\frac{1}{3}x$ = fraction of the job Miah can do in x hours

STEP 3   Set the sum of the two fractions equal to 1 and solve the resulting equation for x.

(**Note:** Step 3 says that the sum of the parts of the job, the fractions, is equal to the whole job, represented by the number 1.)

a. Add the x's. $\frac{1}{2}x + \frac{1}{3}x = 1$

$\frac{1}{2} + \frac{1}{3} = \frac{3}{6} + \frac{2}{6} = \frac{5}{6}$ $\frac{5}{6}x = 1$

b. Multiply $\frac{5}{6}$ by its $\frac{6}{5}\left(\frac{5}{6}\right)x = \frac{6}{5}(1)$

reciprocal of $\frac{6}{5}$ so that

x stands alone. $x = \frac{6}{5} = 1\frac{1}{5}$

ANSWER: $1\frac{1}{5}$ hr = 1 hr 12 min

The steps used to solve the example can be summarized as follows.

STEP 1   Find the fractional part of the job each person can do in 1 hour.

STEP 2   Multiply each fraction from Step 1 by x.

STEP 3   Set the sum of the fractions equal to 1 and solve for x.

Solve.

1. Kami can wash her apartment windows in 1 hour if she works alone. Jeff, working alone, also takes just 1 hour to do the same windows. If they work together, how long will it take them to wash the windows?

2. By herself, Mariah can trim the fruit trees in 2 hours. Paula can do the same job by herself in 4 hours. Working together, how long will it take Mariah and Paula to trim the trees?

3. Working alone, Carrie can paint a room in 4 hours. Amy, working alone, can paint the same room in 3 hours. How long will it take both of them working together to paint the room?

4. Georgi can wax a large truck in 5 hours when he is working alone. Jerry can do the same job alone in 4 hours. How long will it take Georgi and Jerry to wax the truck if they work together?

5. Using the larger tractor, Warren can plow a field in 3 hours. Using the smaller tractor, Eva can plow the same field in 5 hours. How long will it take them to plow the field working together?

Mixture Problems

Do you ever wonder how the prices are determined for different mixtures of nuts or candies? For example, you know that a mixture of cashews and peanuts costs more than peanuts alone but costs less than cashews alone. How is the exact price determined? With algebra, you can answer this question.

EXAMPLE   How many pounds of a $5 per pound mixture of cashews should be mixed with 3 pounds of a $2 per pound mixture of peanuts to obtain a new mixture with a value of $4 per pound?

STEP 1   Let x = the amount of $5 per pound mixture of nuts needed. Put the information you know in a table as follows:

a. Row A is the cost per pound of each mixture.
b. Row B is the number of pounds of each mixture.
c. Row C is the dollar value of each mixture: ($).
    ($) = cost per pound times number of pounds

		Cashews	Peanuts	Mixture
Row A	cost per pound:	$5	$2	$4
× Row B	× number of pounds:	× x	× 3	$x + 3$
Row C	value ($)	5x +	6 =	4(x + 3)

(Row C = Row A × Row B)

Row C gives the equation to solve: $5x + 6 = 4(x + 3)$. The value of the $5 per pound mixture ($5x$) plus the value of the $2 per pound mixture ($2 × 3 = $6) is equal to the value of the $4 per pound mixture $4($x$ + 3).

In words: value of $5 mixture + value of $2 mixture = value of $4 mixture

In symbols: $5x + 6 = 4(x + 3)$

STEP 2   Solve the equation for x. $5x + 6 = 4(x + 3)$

a. Remove the parentheses. $5x + 6 = 4x + 12$
b. Subtract 6 from both sides. $5x = 4x + 12 - 6$
                                                            $5x = 4x + 6$
c. Subtract $4x$ from both sides. $5x - 4x = 6$
                                                                    $x = 6$

ANSWER: 6 pounds of the $5 per pound mixture of cashews should be used.

Solve.

1. How many pounds of a $6 per pound mixture of nuts should be mixed with 5 pounds of a $3 per pound mixture of candy to obtain a mixture of candy and nuts costing $5 per pound?

2. How many pounds of a dried fruit mix costing $3.00 per pound should be mixed with 8 pounds of a nut mixture costing $4.50 per pound to give a fruit-nut mixture costing $4.00 per pound?

3. How many pounds of chocolates costing $2.00 per pound should be mixed with 3 pounds of chocolates costing $1.60 per pound to obtain a mixture of chocolates costing $1.70 per pound?

Temperature: Finding Fahrenheit from Celsius

The temperature formula introduced on page 109 can be rewritten without parentheses so that °F appears on the left side of the equals sign.

$$°F = \frac{9}{5}°C + 32°$$

In this form, you can use the formula to find an equivalent Fahrenheit temperature for a given Celsius temperature.

EXAMPLE   Find the Fahrenheit temperature when the Celsius temperature is 40°.

$$°F = \frac{9}{5}°C + 32$$

        STEP 1 Substitute 40° for °C.

$$°F = \frac{9}{5}(40) + 32$$

        STEP 2 Multiply 40 by $\frac{9}{5}$.

$$\frac{\overset{8}{\cancel{40}}}{1} \times \frac{9}{\underset{1}{\cancel{5}}} = 72$$

        STEP 3 Add.

$$°F = 72 + 32 = \mathbf{104°}$$

ANSWER: 104°F

For problems 1–5, use the temperature formula $°F = \frac{9}{5}°C + 32°$.

1. Water freezes at 0° Celsius. What is the Fahrenheit temperature at which water freezes?

2. Mary likes to be outdoors when the temperature reaches 25°C. What would be the equivalent Fahrenheit temperature?

3. What is the oven temperature in °F when meat is broiling at 180°C?

4. When Raul had the flu, his temperature was 38°C. What was his temperature in °F?

5. A comfortable room temperature is 20°C. What is the equivalent temperature in °F?

Use the temperature formula to complete the Table of Values.
- Fill in the Points to Graph with the coordinates of the three points identified in the completed Table of Values.
- Plot the three points on the graph.
- Connect the points with a straight line and extend the line to the edge of the graph.

6. **Table of Values**

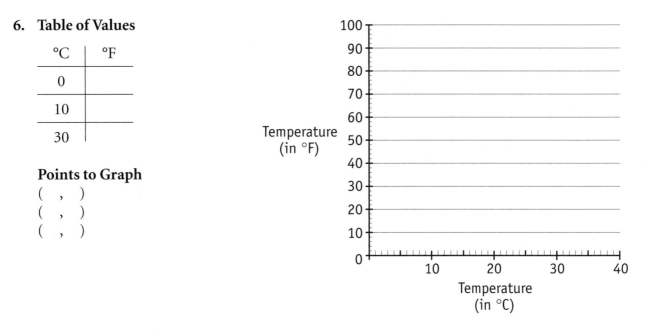

°C	°F
0	
10	
30	

Points to Graph

(,   )
(,   )
(,   )

7. Use the graph to find the approximate temperature, in °F, when the temperature is 28°C.

8. By reading the graph, find the approximate temperature, in °C, when the temperature is 100°F.

9. What is the slope of the temperature line?

Using Algebra in Geometry

Algebra is often used to solve problems in geometry. In the examples below, unknown angles are represented by variables, and equations are used to find the measure of each angle.

From geometry, you know that an angle ($\angle$) is measured in units called **degrees** (°). A right angle ($\llcorner$), formed by perpendicular lines contains 90°. An angle can be labeled with three letters, the second letter being the **vertex,** the point where the two sides meet.

EXAMPLE 1 $\angle ABC$ is a right angle (90°) that is divided into three smaller angles.

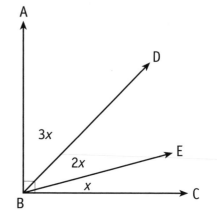

$\angle DBE = 2$ times $\angle EBC$ or $2\angle EBC$ and

$\angle ABD = 3$ times $\angle EBC$ or $3\angle EBC$.

What is the measure of each of these smaller angles?

STEP 1 Let $\angle EBC = x$, $\angle DBE = 2x$, and $\angle ABD = 3x$.

STEP 2 Set the sum of the angles equal to 90° and solve the equation.

$$x + 2x + 3x = 90$$
$$6x = 90$$
$$x = \frac{90}{6} = \mathbf{15}$$
$$2x = 2(15) = \mathbf{30}$$
$$3x = 3(15) = \mathbf{45}$$

ANSWER: $\angle EBC = 15°$, $\angle DBE = 30°$, and $\angle ABD = 45°$

For Example 2, you need to remember that *the sum of the angles in a triangle equals 180°.* Also, an angle in a triangle is often represented by one letter only.

EXAMPLE 2 In $\triangle ABC$, $\angle B = 3\angle A$, and $\angle C = 5\angle A$. What is the measure of each angle?

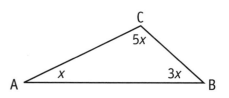

STEP 1 Let $\angle A = x$, $\angle B = 3x$, and $\angle C = 5x$.

STEP 2 Set the sum of the angles equal to 180° and solve the equation.

$$x + 3x + 5x = 180$$
$$9x = 180$$
$$x = \frac{180}{9} = \mathbf{20}$$
$$3x = 3(20) = \mathbf{60}$$
$$5x = 5(20) = \mathbf{100}$$

ANSWER: $\angle A = 20°$, $\angle B = 60°$, and $\angle C = 100°$

For each problem, solve for the angles as indicated.

1. ∠ABC = 90°
   ∠DBE = 3∠EBC
   ∠ABD = 5∠EBC

   What is the measure of each angle?

   ∠EBC =
   ∠DBE =
   ∠ABD =

   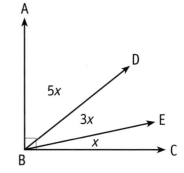

2. ∠D + ∠E + ∠F = 180°
   ∠E = 3∠D
   ∠F = 2∠D

   What is the measure of each angle?

   ∠D =
   ∠E =
   ∠F =

   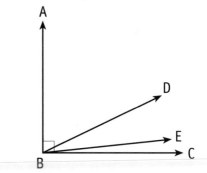

3. ∠ABC = 90°
   ∠DBE = 5∠EBC
   ∠ABD = 9∠EBC

   What is the measure of each angle?

   ∠EBC =
   ∠DBE =
   ∠ABD =

   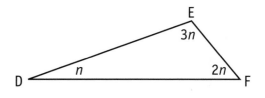

4. ∠A + ∠B + ∠C = 180°
   ∠B = 4∠A
   ∠C = 5∠A

   What is the measure of each angle?

   ∠A =
   ∠B =
   ∠C =

   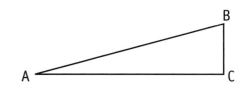

Right Triangles and the Pythagorean Theorem

A **right triangle** is a triangle in which two sides meet at a right angle (90°). The side opposite the right angle is called the **hypotenuse.**

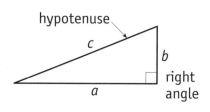

The Greek mathematician Pythagoras discovered that the square of the length of the hypotenuse of a right triangle is equal to the sum of the squares of the lengths of the other two sides. This relation is called the **Pythagorean theorem.**

Using the labels in the triangle above, you can write the Pythagorean theorem.

Pythagorean theorem: $c^2 = a^2 + b^2$

In words: hypotenuse squared = side squared + side squared

EXAMPLE 1 Find the length of the hypotenuse in the triangle at the right.

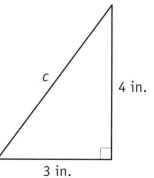

STEP 1 Substitute 3 for a and 4 for b in the Pythagorean theorem.

$$c^2 = a^2 + b^2$$
$$c^2 = 3^2 + 4^2$$
$$= 9 + 16$$
$$= 25$$

STEP 2 Solve for c.

$$c^2 = 25$$
$$c = \sqrt{25}$$
$$c = 5$$

ANSWER: length of hypotenuse = 5 inches

EXAMPLE 2 Find the length of the unlabeled side in the triangle at the right.

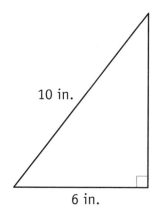

STEP 1 Substitute 10 for c and 6 for a in the Pythagorean theorem.

$$c^2 = a^2 + b^2$$
$$10^2 = 6^2 + b^2$$
$$100 = 36 + b^2$$

STEP 2 Subtract 36 from each side and solve for b.

$$100 - 36 = b^2$$
$$64 = b^2$$
$$\sqrt{64} = b$$
$$8 = b$$

ANSWER: length of unlabeled side = 8 inches

Solve.

1. The two sides of a right triangle measure 6 feet and 8 feet. What is the length of the hypotenuse?

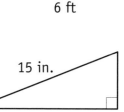

8 ft

6 ft

2. The hypotenuse of a right triangle is 15 inches long. If one side measures 12 inches, what is the length of the other side?

15 in.

12 in.

3. A ladder leans against the side of the house. The length of the ladder is 17 feet. The base of the ladder is 8 feet from the house. How far off the ground is the top of the ladder?

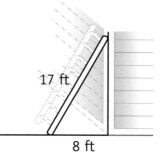

17 ft

8 ft

4. Bill has a garden in the shape of a right triangle. One side measures 5 yards and the other side measures 12 yards. How long is the third side—the side opposite the corner?

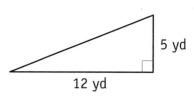

5 yd

12 yd

5. In a right triangle, one side measures 5 feet and the second side measures 7 feet. Find the approximate length of the hypotenuse.

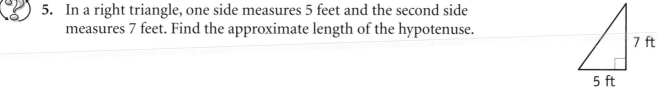

7 ft

5 ft

6. A ship sails 10 miles west and 7 miles north of the harbor. Approximately how far is the ship from the harbor?

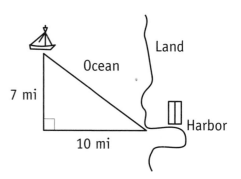

Land

Ocean

7 mi

10 mi

Harbor

Finding the Distance Between Points on a Grid

You may be asked to find the distance between two points on a grid. If the points lie on a horizontal or vertical line, simply find the number of spaces between the points.

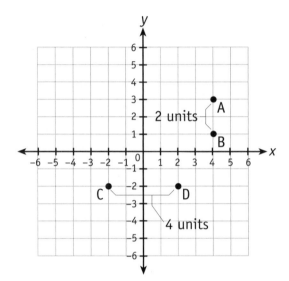

EXAMPLE 1 On the grid at the right, what is the distance between point A and point B?

> **STEP 1** Notice that the points lie on a vertical line parallel to the *y*-axis.
>
> **STEP 2** Count the number of units between the *y*-coordinates.
> 2 units (3 − 1 = 2)

ANSWER: 2 units

EXAMPLE 2 Find the distance between point C and point D.

> **STEP 1** Notice that the points lie on a line parallel to the *x*-axis.
>
> **STEP 2** Count the number of units between the *x*-coordinates.
> 4 units [2 − (−2) = 2 + 2 = 4]

ANSWER: 4 units

When the points do not lie on a horizontal or vertical line, they lie on a line called a **diagonal.** The Pythagorean theorem (previous lesson) is used to find the distance between two points that lie on a diagonal.

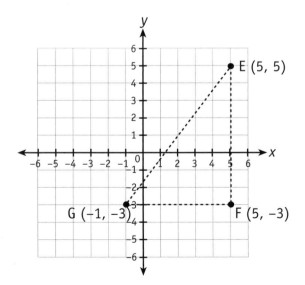

EXAMPLE 3 Find the distance between point E and point G.

> **STEP 1** Draw three lines to connect points E, F, and G.
>
> **STEP 2** Find the difference between the *x*-coordinates and the difference between the *y*-coordinates. (Subtract the lesser from the greater.)
>
> *x*-coordinates: 5 − (−1) = 5 + 1 = **6**
> *y*-coordinates: 5 − (−3) = 5 + 3 = **8**
> 6 and 8 are the lengths of the two shorter sides.

STEP 3 Substitute the values 6 and 8 into the Pythagorean theorem and solve for EG—the diagonal from point E to point G and the hypotenuse of right triangle EFG.

$$c^2 = a^2 + b^2$$
$$(EG)^2 = 6^2 + 8^2$$
$$= 36 + 64$$
$$= 100$$
$$EG = \sqrt{100}$$
$$\mathbf{EG = 10}$$

ANSWER: 10 units

· ·

Find the distance between the points on the grid.

1. points A and B

2. points A and C

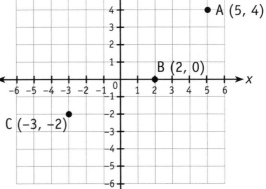

Find the distance between the points on the graphed lines.

3. points A and B

4. points B and C

5. points A and C

6. points A and D

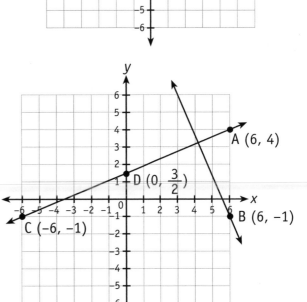

Absolute Value

Absolute value is the positive value of a number or expression regardless of its sign. The symbol for absolute value is | |.

Number	Absolute Value	Read as
7	$\lvert 7 \rvert = 7$	The absolute value of 7 is 7.
–7	$\lvert -7 \rvert = 7$	The absolute value of –7 is 7.

EXAMPLE 1 If $\lvert x \rvert = 3$, what are the possible values of x?

Notice that $\lvert 3 \rvert = 3$ and $\lvert -3 \rvert = 3$.

ANSWER: $x = 3$ and $x = -3$

On a number line, the distance between the numbers x and a is equal to $\lvert x - a \rvert$. An unknown distance may be written as an **absolute-value equation.**

EXAMPLE 2 If $\lvert n - 3 \rvert = 2$, what are the possible values of n?

The same question can be asked this way. "On a number line, what values of n are each 2 units away from 3?"

1 is 2 units to the left of 3. 5 is 2 units to the right of 3.
$\lvert 1 - 3 \rvert = \lvert -2 \rvert = 2$ $\lvert 5 - 3 \rvert = \lvert 2 \rvert = 2$

ANSWER: $n = 1$ and $n = 5$

To tell a range of values, an **absolute-value inequality** such as $\lvert x - 1 \rvert < 6$ may be used.

EXAMPLE 3 For what values of x is $\lvert x - 1 \rvert < 6$ true?

$\lvert x - 1 \rvert < 6$ is true for all values of x that are *less than 6 units away from 1*. The graph shows this range of values.

ANSWER:

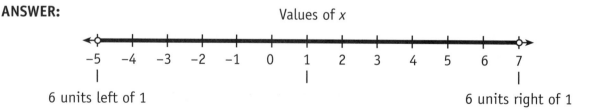

Values of x

6 units left of 1   6 units right of 1

(**Note:** $\lvert x - 1 \rvert < 6$ can also be written as an *and inequality* (see page 123). As an *and inequality,* the values of x are written $-5 < x < 7$.)

Write the absolute value of each number.

1. $|-4| =$ $|0| =$ $|8.5| =$ $\left|-\frac{1}{2}\right| =$ $|+9| =$

Find the value of each expression. The first one is done as an example.

2. $|7-9| = \left|-2\right| = 2$ $|3-12| =$ $|-1-6| =$

Find the solutions to each absolute-value equation.

3. $|x| = 5$ $|n| = 3$ $|r| = 10$ $|w| = 0$ $|z| = 100$

4. $|x-2| = 3$ $|n-5| = 1$ $|r-7| = 15$

Graph each absolute-value inequality.

5. $|x-4| < 5$

Values of x

```
◄──┼───┼───┼───┼───┼───┼───┼───┼───┼───┼───┼───┼──►
   -2  -1   0   1   2   3   4   5   6   7   8   9  10
```

6. $|n-1| \le 3$

Values of n

```
◄──┼───┼───┼───┼───┼───┼───┼───┼───┼───┼───┼───┼──►
   -6  -5  -4  -3  -2  -1   0   1   2   3   4   5   6
```

7. Write an absolute-value inequality to represent the values of r.

Values of r

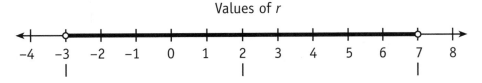

Answer Key

Pages 1–6

1. $-1 + -6 = -7$

2. $5 + -3 = 2$

3. 11

4. −6

5. −30

6. 2

7. 16

8. −27

9. 6^6

10. 7^3

11. 8

12. $\frac{5}{7}$

13. $3(n + 6)$

14. $20\ [24 - 4]$

15. $16\ [8(2)]$

16. $100°C\ [\frac{5}{9}(180)]$

17. c. $x - 8 = 21$

18. a. $x = 17$ **b.** $y = 22$

19. a. $n = 6$ **b.** $n = 36$

20. a. $x = 9$ **b.** $s = 48$

21. a. $y = 8$ **b.** $n = 8$

22. a. $x = 13$ **b.** $b = 5$

23. 21 and 28

24. $5\frac{1}{3}$ cups $\left(\frac{n}{16} = \frac{2}{6}\right)$

25. $x = 5$ or $x = -5$

26.

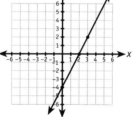

27. slope = $\frac{4}{3}$

28.

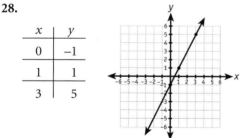

x	y
0	−1
1	1
3	5

29. $n > -3$

30. $n \geq 2$

31. $x < 7$

32.

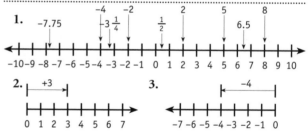

Values of x

33. $-3x^2$ and $5x^2$

34. $3x^2 - x$

35. $a^2 - 5b$

36. $6n^3 + 8n^2 - 4n$

37. $7x^2 + 3x$

38. $5\sqrt{3}$

39. $6x^2\sqrt{y}$

40. $2(2x^2 + 5)$

41. $4z^2(3z + 4)$

42. $3r^2 + 4$

43. 7

Page 10

1. $+8, +3, +12, +7$

2. $-5, -4, -19, -8$

3. $+2\frac{1}{2}, +4.75, -5\frac{2}{3}, -\frac{1}{4}$

4. positive, negative, positive, negative

5. positive, positive, negative, negative

Page 11

1.

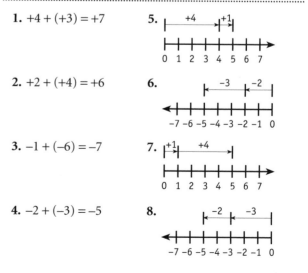

2.

3.

Pages 13 and 15

1. $+4 + (+3) = +7$

2. $+2 + (+4) = +6$

3. $-1 + (-6) = -7$

4. $-2 + (-3) = -5$

5.

6.

7.

8.

For problems 9–11, a + sign does not need to be placed before answers that are positive.

9. 11, 11, 19, 16

10. −7, −7, −7, −12

11. −21, −21, −25, −41

12. 5 + (−3) = +2   **13.** 5 + (−5) = 0

14. 1 + (−3) = −2   **15.** −2 + 4 = +2

16.   **17.**

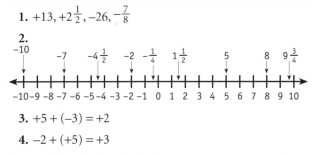

18. 3, −3, 0, −1

19. 1, −3, 4, −7

20. 0, −5, 5, 9

Page 17

1. b. 5 − (−4)   **2. a.** 2 − (−3)

For problems 3–6, a + sign does not need to be placed before answers that are positive.

3. 6, 5, 4, 8

4. −4, 5, −4, 6

5. 5, 10, 18, 23

6. −6, −12, −1, −6

7. 13° [7 − (−6)]

8. 111° below freezing [32 − (−79)]

9. 20,600 feet [20,320 − (−280)]

10. 14,775 feet [14,495 − (−280)]

Page 19

A + sign does not need to be placed before answers that are positive.

1. 18, −35, 32, 54

2. 56, 20, −48, −12

3. 36, 28, 42, 18

4. −40, −9, −36, −30

5. 45, −54, 21, −24

6. 30, 16, 48

7. −24, −18, −36

8. 12, −40, 48

9. −16, 30, 8

Page 21

A + sign does not need to be placed before answers that are positive.

1. −6, −5, 6, 7

2. 13, −34, 18, −14

3. −3, −9, 9, 8

4. −25, −14, 30, −24

5. −2, −2, 6, −4

6. −14, −15, 12, −25

7. $-\frac{2}{3}, -\frac{3}{4}, -\frac{1}{2}, -\frac{2}{5}$

8. $\frac{1}{2}, \frac{2}{5}, \frac{3}{4}, \frac{2}{3}$

9. $-\frac{1}{5}, -\frac{2}{5}, -\frac{7}{20}, -\frac{2}{3}$

Pages 22–23

1. $+13, +2\frac{1}{2}, -26, -\frac{7}{8}$

2.

3. +5 + (−3) = +2

4. −2 + (+5) = +3

For problems 5–8 and 10–16, a + sign does not need to be placed before answers that are positive.

5. 5, 3, 0, 9

6. −12, −12, 3, −4

7. 5, 0, −3, 19

8. 2, −13, −8, −6

9. 23° [17 − (−6)]

10. −27, −28, −18, −16

11. 30, 15, 32, 54

12. −24, 36, 18

13. −5, −7, 8, 8

14. −4, 8, −7, −6

15. −6, −2, 3, −9

16. $\frac{7}{8}, -\frac{4}{5}, -\frac{3}{4}, \frac{2}{3}$

Pages 25–26

1. $(-1)^2$, −1 to the second power or −1 squared

2. 5^2, 5 squared or 5 to the second power

3. $\left(\frac{1}{2}\right)^2$, $\frac{1}{2}$ squared or $\frac{1}{2}$ to the second power

4. 6^2, 6 squared or 6 to the second power

5. 7^3, 7 cubed or 7 to the third power

6. $(-3)^3$, -3 cubed or -3 to the third power

7. 6^3, 6 cubed or 6 to the third power

8. $(-\frac{3}{4})^3$, $-\frac{3}{4}$ cubed or $-\frac{3}{4}$ to the third power

9. 2^4, 2 to the fourth power

10. 5^4, 5 to the fourth power

11. $(-4)^4$, -4 to the fourth power

12. $(\frac{1}{4})^4$, $\frac{1}{4}$ to the fourth power

13. 25, 4, 81, 36

14. 125, 81, -1000, 81

15. $\frac{4}{9}$, $-\frac{64}{125}$, $\frac{1}{16}$, $\frac{27}{64}$

16. 61, 80, 5

17. 14, 56, 21

18. $\frac{3}{4}$, $\frac{1}{16}$, $1\frac{1}{16}$

Page 27

1. 128, 432, -200

2. $3\frac{1}{8}$, 5, $\frac{4}{27}$

3. $2\frac{27}{49}$, $1\frac{32}{49}$, -4

4. $3\frac{5}{9}$, $2\frac{13}{18}$, $\frac{5}{18}$

Pages 28–29

1. 2^7, 3^4, 4^7

2. 5^3, 7^4, 4^4

3. 6^3, 2^5, 8^7

4. $4^5 \cdot 3^4$, $5^{12} \cdot 2^3$, $7^6 \cdot 5^3$

5. 5^2, 2^4, 8^2

6. $7^1 = 7$, $8^1 = 8$, $5^1 = 5$

7. 4^2, 8^3, $7^1 = 7$

8. $6^3 \cdot 4^2$, $10^2 \cdot 5^4$, $9^2 \cdot 7^3$

Page 30

1. $\frac{1}{7 \times 7}$, 7 to the minus 2nd power, $\frac{1}{49}$

2. $\frac{1}{10 \times 10 \times 10}$, 10 to the minus 3rd power, $\frac{1}{1,000}$

3. $\frac{1}{3 \times 3 \times 3 \times 3}$, 3 to the minus 4th power, $\frac{1}{81}$

4. 3^2, 4^0 or 1, 10^{-1}

5. 6^3, 2^6, 5^4

Page 31

1. 630; 8,750,000; 90,000,000

2. 0.051; 0.000085; 0.000007

3. 5×10^3, 3×10^4, 7.5×10^7

4. 3×10^{-2}, 7.5×10^{-3}, 1.25×10^{-5}

5. 240,000 miles

6. 1.2×10^{-3} inches

Page 33

1. ±13, ±5, ±9

2. ±2, ±11, ±6

3. ±7, ±1, ±15

4. ±12, ±8, ±3

5. ±10, ±4, ±14

6. $\pm\frac{2}{3}$, $\pm\frac{3}{8}$, $\pm\frac{5}{6}$

7. $\pm\frac{1}{2}$, $\pm\frac{7}{9}$, $\pm\frac{11}{12}$

8. $\pm\frac{9}{4}$, $\pm\frac{1}{14}$, $\pm\frac{10}{11}$

Page 35

1. 6.5

2. 4.4

3. 8.8

4. 9.5

5. 5.3

6. 7.1

7. 11.7

8. 14.4

Pages 36–37

1. 3 is the base, 4 is the exponent

2. $(-4)^4$

3. 5^3

4. $(-\frac{3}{5})^2$

5. 8^2

6. $(\frac{2}{3})^3$

7. 9^4

8. 81

9. -125

10. 81

11. 1

12. 7

13. $\frac{4}{9}$

14. $17(25 - 9 + 1)$

15. $1\frac{1}{2}$

16. $7\frac{1}{9}$

17. $\frac{9}{10}$

18. 5^9

19. 8^3

20. $\frac{1}{64}$

21. 2

22. $\frac{1}{1,000}$

23. 45,000

24. 0.0023

25. 7.2×10^5

26. 6×10^{-3}

27. ±6

28. $\pm\frac{7}{8}$

29. $\sqrt{56} \approx 7.5$

30. $\sqrt{83} \approx 9.1$

Pages 38–39

For problems 1–12, answers may vary.

1. $n - 8$
2. $\frac{s}{5}$
3. $y + 12$
4. $w - 4$
5. $7x$
6. $19 + r$
7. 5 times x
8. 4 plus n
9. 7 minus n
10. m divided by 2
11. a divided by 5
12. s subtract 3
13. e
14. h
15. a
16. f
17. g
18. b
19. c
20. d

Page 40

Answers may vary.

1. a. $3 + 4n$ b. $2z + 4$ c. $5 + x$
2. a. $3s - 1.2$ b. $x - 16$ c. $4y - 8$
3. a. $2z + 7$ b. $3s + 6$ c. $6x + 8$
4. a. $\frac{3y}{2}$ b. $\frac{6p}{4}$ c. $\frac{r}{3}$

Page 41

1. $2x + 7$
2. $13y + 12$
3. $3s^2 + s + 9$
4. $10x + 8$
5. $4y^2 + 6$
6. $6n^2 + 8n + 2$

Pages 42–43

1. c. $h + 144$
2. a. $\frac{t}{3}$
3. d. $x - \$2.95$
4. b. $\frac{(n-5)}{3}$
5. d. $n - 0.05n$
6. c. $g^2 - 6$
7. a. $6n^2 + 3$
8. c. $(n - 3)^2$

For problems 9–14, answers may vary.

9. $h + 154$
10. $\frac{n}{5}$
11. $0.3p$
12. $m - 0.25m$
13. $3n^2 - 4$
14. $2b^2 - 12b$

Pages 45–46

1. $5 (-4 + 9)$ $-9 (3 - 12)$ $4 (9 + -5)$
2. $20 (5 \cdot 4)$ $-14 (7 \cdot -2)$ $16 (\frac{2}{3} \cdot 24)$
3. $7 (\frac{35}{7})$ $-6 (\frac{-18}{3})$ $5 (\frac{-10}{-2})$
4. $-8 (-2 \cdot 4)$ $6 (-3 \cdot -2)$ $-8 (4 \cdot -2)$
5. $57 (12 + 45)$ $0 (12 - 12)$ $13 (25 - 12)$
6. $9 (12 - 3)$ $-39 [-36 + -3]$ $18 [2 + 24 - 8]$
7. $27 [3(9)]$ $8 [-4(-2)]$ $-7 [7(-1)]$
8. $27 (18 + 9)$ $100 (36 + 64)$ $21 (3 + 18)$

9. $10 \left(\frac{50}{5}\right)$ $12 \left(\frac{48}{4}\right)$
10. $-20 \left(\frac{-100}{5}\right)$ $\frac{1}{3} \left(\frac{4}{12}\right)$

Pages 47–48

1. 22 feet $[2(7 + 4)]$
2. 28 feet $(6 + 9 + 13)$
3. 44 inches $[2(\frac{22}{7})7]$
4. 84 square feet $(12 \cdot 7)$
5. 42 square inches $(\frac{1}{2} \cdot 14 \cdot 6)$
6. 616 square inches $(\frac{22}{7} \cdot 14 \cdot 14)$
7. 720 cubic inches $(16 \cdot 5 \cdot 9)$
8. 88 cubic feet $(\frac{22}{7} \cdot 2 \cdot 2 \cdot 7)$

Page 49

1. 18 feet $(5 + 4 + 5 + 4)$
2. 466 yards $(150 + 125 + 191)$
3. about 176 feet $[2 \cdot (\frac{22}{7}) \cdot 28]$
4. 180 square feet $(15 \cdot 12)$
5. 18 square yards $(\frac{1}{2} \cdot 12 \cdot 3)$
6. about 1,386 square meters $(\frac{22}{7} \cdot 21 \cdot 21)$
7. 756 cubic feet $(18 \cdot 6 \cdot 7)$

Pages 50–51

For problems 1–16, answers may vary.

1. $x - 7$
2. $\frac{z}{9}$
3. $y + 19$
4. $-4n$
5. $6(x + 4)$
6. $2(y - 5)^2$
7. 12 plus y
8. x minus 9
9. 15 times a
10. y divided by 12
11. 4 times s plus 8
12. -9 times the quantity 2 times x plus 4
13. $6s$
14. $m - \$12.49$
15. $\frac{(n-2)}{6}$
16. $n - 0.4n$
17. $-2 [13 + (-15)]$
18. $34 (42 - 8)$
19. $7 (-2 + 9)$
20. $60 (4 \cdot 5 \cdot 3)$
21. $-9 [-3(3)]$
22. $-16 (8 - 24)$
23. $12 [(-2)(-6)]$
24. $4 \left(\frac{32}{8}\right)$
25. about 283 cubic feet $(\frac{22}{7} \cdot 3 \cdot 3 \cdot 10)$
26. $\$74.25 (\$550 \cdot 0.09 \cdot 1.5)$
27. 180 square feet $(15 \cdot 12)$
28. $40°C [\frac{5}{9}(72)]$

Pages 52–53

1. b. $9n = 14$ **2. c.** $\frac{n}{6} = 50$

For problems 3–10, answers may vary.

3. $y + 9 = 26$ **7.** $5x = 60$

4. $x + 14 = 21$ **8.** $4y = 24$

5. $t - 13 = 15$ **9.** $\frac{n}{3} = 7$

6. $x - 21 = 43$ **10.** $\frac{s}{6} = 2$

Page 54

1. No **5.** Yes

2. Yes **6.** No

3. Yes **7.** Yes

4. No **8.** Yes

Page 56

1. $x = 3, x = 5, y = 6, n = 14$

2. $y = \$13, s = \$5, x = \$23, z = 18¢$

3. $x = 3\frac{1}{2}, y = 2\frac{1}{3}, c = \frac{1}{2}, r = 4\frac{1}{4}$

4. $x = 3.6, n = 3.3, z = 8.5, y = 4.75$

Page 57

1. $x = 20, x = 10, y = 16, z = 47$

2. $y = \$17, x = \$26, y = \$37, n = 36¢$

3. $x = 6\frac{2}{3}, r = 3\frac{5}{8}, y = 2, v = 8$

4. $m = 7.7, x = 8.1, n = 17.25, y = 11.75$

Page 58

1. c. $n - 18 = 14, n = 32$

2. a. $m - \$5.75 = \$2.15, m = \$7.90$

3. a. $p + 0.25(\$40) = \$40, p = \$30$

4. c. $t + 15°F = 87°F, t = 72°F$

For problems 5 and 6, equations may vary.

5. $p - \$14.95 = \$34.00, p = \$48.95$

6. $n - 14 = 19, n = 33$

Page 59

1. $x = 5, n = 6, y = 9, m = 7$

2. $y = -6, x = -9, b = -13, y = -11$

3. $x = \$3, n = \$7, z = \$21, y = 11¢$

4. $y = 2.4, y = 6.5, x = 5.4, n = 7.5$

Page 60

1. $x = 20, y = 27, c = 88, n = 48$

2. $y = \$18, x = \$68, n = \$36, t = \$18.00$

3. $x = 7, y = 7, n = 15, r = 13.5$

Page 61

1. c. $14n = 168, n = 12$

2. b. $0.05w = 9, w = 180$

3. a. $\frac{n}{4} = 7, n = 28$

4. b. $s = \frac{\$16.45}{3}, s \approx \$5.48$

For problems 5 and 6, equations may vary.

5. $h = 5 \times 6.5, h = 32.5$

6. $\frac{s}{8} = 4, s = 32$

Page 62

1. $x = 8, z = 12, a = 28, n = 12$

2. $n = -\frac{7}{4}, s = -1, x = -\frac{8}{9}, y = -\frac{35}{16}$

3. $z = \$5.25, x = \$1.60, c = \$17.60, r = \$2.10$

Page 63

1. b. $\frac{2}{3}x = 36, x = 54$

2. a. $\frac{2}{5}i = \$350, i = \$875$

3. c. $\frac{3}{4}n = 18, n = 24$

4. b. $\frac{4}{5}r = \$7.75, r \approx \$9.69$

For problems 5 and 6, equations may vary.

5. $\frac{3}{5}h = 21, h = 35$

6. $\frac{5}{8}c = 145, c = 232$

Pages 64–67

1. $x = 2$ **12.** $a = -24$ **23.** $x = 25$

2. $z = 13$ **13.** $x = -15$ **24.** $s = 0$

3. $a = -3$ **14.** $y = 36$ **25.** $x = \frac{2}{3}$

4. $y = 19$ **15.** $a = -8$ **26.** $r = 6.75$

5. $x = 25$ **16.** $y = 17$ **27.** $x = \frac{1}{15}$

6. $b = -4$ **17.** $n = 35$ **28.** $k = 3.12$

7. $z = 8$ **18.** $q = 24$ **29.** $x = 1\frac{1}{5}$

8. $x = -8$ **19.** $y = 35$ **30.** $n = 2$

9. $y = -11$ **20.** $n = -7$ **31.** $z = 42$

10. $x = 27$ **21.** $x = -3$ **32.** $y = 11.25$

11. $y = 56$ **22.** $p = 7$ **33.** $x = 4\frac{1}{2}$

For problems 34–39, answers to parts a and b may vary.

34. a. m = money Zeta had
b. $m - \$4.75 = \$2.40$
c. $m = \$7.15$

35. a. c = cookies in each box
b. $\frac{c}{5} = 6$
c. $c = 30$

36. a. m = money Ephran paid clerk
b. $m - \$7.89 = \$2.11$
c. $m = \$10.00$

37. a. n = number of students in class
b. $\frac{3}{4}n = 18$
c. $n = 24$

38. a. n = number of teachers at Grant School
b. $\frac{1}{6}n = 4$
c. $n = 24$

39. a. m = amount Jocelyn earned during May
b. $\frac{2}{3}m = \$90$
c. $m = \$135$

For problems 40–47, equations may vary.

40. $x + 7 = 17, x = 10$
41. $n + 13 = 21, n = 8$
42. $5y = 45, y = 9$
43. $\frac{r}{3} = 4, r = 12$
44. $x + 19 = 28, x = 9$
45. $y + 32 = 62, y = 30$
46. $16w = 64, w = 4$
47. $5l = 80, l = 16$

Page 69

1. $y = 3, x = 3, z = -4$
2. $x = 4, z = 8, a = -7$
3. $y = -8, z = 2, x = -30$

Page 71

1. $x = 6, y = 7, z = 5$
2. $y = 13, a = 5, x = 6$
3. $y = 15, z = -4, n = -24$
4. $x = 1, a = 3, z = 5$
5. $x = 9, z = 21, n = 8$

Pages 72–73

1. $x = 5, y = 4, z = 4$
2. $a = 12, y = 8, x = 5$

3. $x = 4, z = 2, y = 26$
4. $y = 48, a = 50, x = 18$
5. $z = 4, y = 7, a = 2$
6. $a = 10, x = 12, y = 5$
7. $b = 3, z = 2, y = 6$

Pages 74–75

1. $x = 6, y = 3, z = 9$
2. $z = 3, a = 3, x = -3$
3. $x = 8, b = 8, y = 2$
4. $m = 8, x = 2\frac{1}{2}, y = -4$
5. $x = 4, y = 3, z = -9$
6. $a = -10, b = 2, z = 6$
7. $y = 3, x = -5, n = 3$

Pages 76–77

1. b. $8n + 9 = 73, n = 8$
2. c. $5w - 7 = 3w + 19, w = 13$
3. a. $t + 2t = \$225, t = \$75$
4. c. $\$18a + \$400 = \$886, a = 27$

For problems 5–12, equations may vary.

5. $3n = n + 12, n = 6$
6. $\frac{2}{3}m + \frac{1}{6}m = 25, m = 30$
7. $p + 0.06p = \$360.40, p = \$340.00$
8. $s + \$50(6) = 3s, s = \$150$
9. $4x + 15 = 27, x = 3$
10. $3n + 16 = 5n + 10, n = 3$
11. $3r = r + 4 + 4 + 2, r = 5$
12. $4y = 2y + 24, y = 12$
13. $x = 8$ $(2x + 6 + 2x + 4x - 8 = 62,$ or $8x = 64)$
14. $x = 16$ $(x + 3x - 16 + 2x + 10 = 90,$ or $6x = 96)$

Page 79

1. $x = 6, n = 7, b = 5$
2. $y = 11, a = 11, b = 4$
3. $x = 10, m = 7, y = -2$
4. $y = 30, z = 21, a = 13$

Page 81

1. b. $3(n + 1) = 2(n + 4), n = 5$
2. c. $m + (m - \$10) + 2(m - \$10) = \$310, m = \$85$
3. $4(n + 2) = 3(n + 5), n = 7$
4. $\frac{2}{3}(y - 3) = \frac{1}{3}y, y = 6$

5. $s + (s - \$0.45) + 2(s - \$0.45) = \$9.25$, $s = \$2.65$
Sami owes: $2.65 (s)
Ben owes: $2.20 (s - $0.45)
Louis owes: $4.40 [2(s - $0.45)]

6. $x + (x + 1) + (x + 2) = 54$
1st integer: 17 (x)
2nd integer: 18 (x + 1)
3rd integer: 19 (x + 2)

Pages 82–85

1. $x = 16$ **11.** $n = 23$ **21.** $r = 3$

2. $n = 4$ **12.** $y = 17$ **22.** $x = -6$

3. $r = 72$ **13.** $b = 4$ **23.** $y = 10$

4. $y = 11$ **14.** $z = 16$ **24.** $n = -2$

5. $x = 24$ **15.** $y = 4$ **25.** $y = -6$

6. $n = 24$ **16.** $w = 13$ **26.** $z = 4$

7. $n = 5$ **17.** $x = 5$ **27.** $x = 44$

8. $x = 9$ **18.** $n = 7$ **28.** $p = 4$

9. $y = 30$ **19.** $x = 5$ **29.** $n = 6$

10. $x = 5$ **20.** $s = 5$ **30.** $x = 14$

31. b. $6m - 4 = 26$, $m = 5$

32. c. $n + 8(5) = 2n$, $n = 40$

33. $5(x - 1) = 3(x + 9)$, $x = 16$

34. $4n - 3(n + 2) = 5$, $n = 11$

35. $c + 24(5) = 3c$, $c = 60$

36. $h + (h - 8) + 2(h - 8) = 72$
Alyce: 24 hours (h)
Jon: 16 hours (h - 8)
Kim: 32 hours [2(h - 8)]

37. $x + (x + 2) + (x + 4) = 66$
1st integer: 20 (x)
2nd integer: 22 (x + 2)
3rd integer: 24 (x + 4)

38. $(3x - 2) + (2x + 4) + (3x - 3) = 89$, or $8x = 90$, $x = 11.25$

For problems 39–42, equations may vary.

39. $3x + 25 = 33$, $x = 2\frac{2}{3}$

40. $5y + 9 = 4y + 12$, $y = 3$

41. $3d = d + 9 + 6$, $d = 7\frac{1}{2}$

42. $3n + 6 = 7n - 26$, $n = 8$

Page 86

1. $\frac{4}{3}\left(\frac{16}{12}\right)$ **4.** $\frac{2}{7}\left(\frac{4}{14}\right)$

2. $\frac{5}{16}\left(\frac{25}{80}\right)$ **5.** $\frac{5}{17}\left(\frac{100}{340}\right)$

3. a. $\frac{3}{2}\left(\frac{12}{8}\right)$

   b. $\frac{3}{5}\left(\frac{12}{20}\right)$

Page 87

For problems 1–4, equations may vary.

1. $3x + 2x = 75$, $x = 15$
1st number = 45 (3x)
2nd number = 30 (2x)

2. $5x - 2x = 18$, $x = 6$
1st number = 30 (5x)
2nd number = 12 (2x)

3. $\$4x + \$6x = \$740$, $x = \$74$
Alan's earnings = $296 ($4x)
Shannon's earnings = $444 ($6x)

4. $7x + 4x + 7x + 4x = 176$, $x = 8$
length = 56 feet (7x)
width = 32 feet (4x)

Page 89

1. No, Yes, No

2. Yes, No, Yes

3. $16n = 16$, $4n = 36$, $20x = 40$, $12x = 30$

4. $5x = 90$, $32n = 96$, $3x = 48$, $12n = 144$

5. $n = 1$, $x = 2$, $x = 3$, $n = 2$

6. $n = 9$, $x = 12$, $n = 8$, $x = 3$

Pages 90–91

1. b. $\frac{75}{2} = \frac{280}{n}$ **3. a.** $\frac{450}{8} = \frac{225}{t}$

2. c. $\frac{2}{3} = \frac{p}{19}$ **4. c.** $\frac{2}{5} = \frac{n}{280}$

For problems 5–10, variables and proportions may vary. However, there is only one correct answer.

5. $c = $ cost
$\frac{c}{12} = \frac{\$1.74}{8}$
$c = \$2.61$

6. $d = $ distance
$\frac{d}{16} = \frac{102}{3}$
$d = 544$ miles

7. $n = $ quarts of brown
$\frac{n}{15} = \frac{2}{5}$
$n = 6$ quarts

8. $n = $ number of centimeters
$\frac{n}{2} = \frac{12.7}{5}$
$n = 5.08$ centimeters

9. $m = $ additional earnings
$\frac{m}{20} = \frac{\$55.40}{8}$
$m = \$138.50$

10. $d = $ drops of hardener
$\frac{d}{28} = \frac{3}{7}$
$d = 12$ drops

Page 93

1.
x	y
0	4
1	5
2	6
3	7

4.
s	r
−2	2
0	3
2	4
4	5

2.
x	y
1	1
2	4
3	7
4	10

5.
w	P
6	18
10	26
15	36
20	46

3.
b	a
−1	5
0	3
1	1
2	−1

6.
∠N	∠M
25°	65°
45°	45°
62°	28°
78°	12°

Page 95

1. $x = -2, y = -1$
2. $x = -5, y = -9$
3. $x = 1, y = 2$
4. $x = 2, y = 7$
5. $x = -3, y = 3$
6. $x = 4, y = 1$

Page 97

1. $y = 9, y = 8, y = 12$
2. $y = 7, y = 2, y = 37$
3. $x = \pm 5, x = \pm 4, x = \pm 6$
4. $x = \pm 7, x = \pm 8, x = \pm 6$
5. $x = \pm 3, x = \pm 5, x = \pm 13$

Pages 98–99

1. **a.** $\frac{5}{2} \left(\frac{25}{10} \right)$ **b.** $\frac{5}{7} \left(\frac{25}{35} \right)$

2. 1st number = 27 $(3x)$
   2nd number = 36 $(4x)$
   $(3x + 4x = 63)$

3. length = 66 yards $(3x)$
   width = 22 yards (x)
   $[2(3x) + 2(x) = 176$
   or $6x + 2x = 176]$

4. $n = 6$ 5. $y = 2$ 6. $x = 32$

For problems 7–8, proportions may vary.

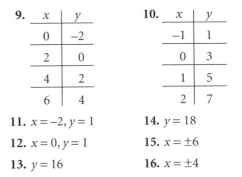

7. $79.20 (m)$
   $\frac{m}{12} = \frac{\$46.20}{7}$ where m = amount that can be earned in 12 hours

8. 24
   $\frac{3}{5} = \frac{n}{40}$ where n = number of Jefferson's men teachers

9.
x	y
0	−2
2	0
4	2
6	4

10.
x	y
−1	1
0	3
1	5
2	7

11. $x = -2, y = 1$
12. $x = 0, y = 1$
13. $y = 16$
14. $y = 18$
15. $x = \pm 6$
16. $x = \pm 4$

Page 101

1. **a.** $(4, -2)$ **b.** $(-3, 5)$ **c.** $(2, 6)$ **d.** $(-4, -7)$

2. **a.** $x = 2, y = -8$
   b. $x = -1, y = 6$
   c. $x = 0, y = 5$
   d. $x = -4, y = 0$

3. A = $(-6, 5)$; B = $(-3, 3)$; C = $(-2, 0)$; D = $(2, 4)$
   E = $(0, 2)$; F = $(6, 2)$; G = $(3, 0)$; H = $(5, -2)$
   I = $(1, -3)$; J = $(0, -5)$; K = $(-7, -5)$; L = $(-3, -3)$

Page 102

1.
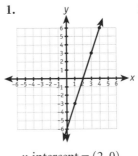

2.
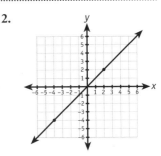

Page 103

1. 2.
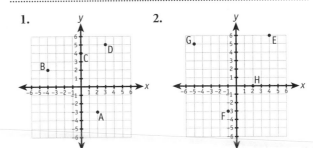

x-intercept = $(2, 0)$ x-intercept = $(0, 0)$
y-intercept = $(0, -6)$ y-intercept = $(0, 0)$

Page 105

1. Line A: zero; Line B: negative
   Line C: undefined: Line D: positive

2. Slope of line E is 2.
   x-intercept $= (2, 0)$
   y-intercept $= (0, -4)$

3. Slope of line F is $-\frac{4}{3}$.
   x-intercept $= (1\frac{1}{2}, 0)$
   y-intercept $= (0, 2)$

4. **a.** $5 \left(\frac{6-1}{4-3} = \frac{5}{1} \right)$

   b. $2 \left(\frac{4-2}{1-0} = \frac{2}{1} \right)$

   c. $-1 \left(\frac{5-2}{-2-1} = \frac{3}{-3} \right)$

   d. $\frac{2}{3} \left(\frac{2-0}{3-0} = \frac{2}{3} \right)$

Pages 106–107

1.

x	y
0	−2
2	0
5	3

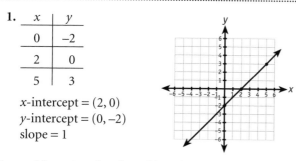

x-intercept $= (2, 0)$
y-intercept $= (0, -2)$
slope $= 1$

For problems 2 and 3, the Tables of Values will vary from the sample answers given. The graphed lines, intercept values, and slope should be the same as those given.

2.

x	y
0	2
2	4
4	6

x-intercept $= (-2, 0)$
y-intercept $= (0, 2)$
slope $= 1$

3.

x	y
0	−4
−1	−2
1	−6

x-intercept $= (-2, 0)$
y-intercept $= (0, -4)$
slope $= -2$

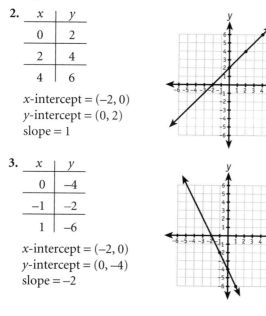

4.

x	y
0	−1
2	$\frac{1}{2}$
4	2

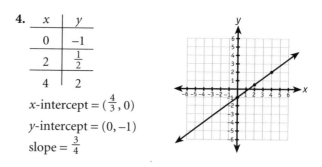

x-intercept $= (\frac{4}{3}, 0)$
y-intercept $= (0, -1)$
slope $= \frac{3}{4}$

Page 108

For problem 1, Table of Values will vary.

1.

r	C
0	0
4	25.12
10	62.8
15	94.2

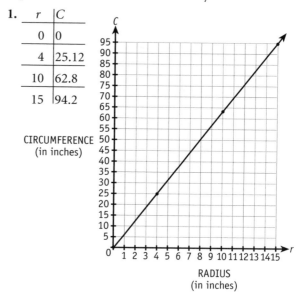

For problems 2–5, estimates will vary.

2. approximately 41 inches

3. approximately 7.5 inches

4. approximately 6.25

Page 109

For problem 1, Table of Values will vary.

1.

°F	°C
10°	−12.2°
32°	0°
50°	10°
100°	≈37.8°

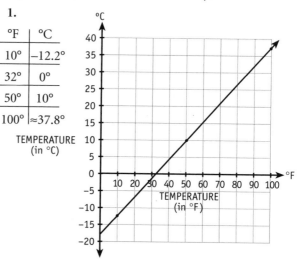

For problems 2–5, estimates will vary.

2. approximately 18°C

3. approximately 32°F

4. approximately $\frac{1}{2}$ (exact slope is $\frac{5}{9}$)

Pages 110–111

1. $y = 2x + 1$ $y = -x + 4$

x	y
−2	−3
0	1
2	5

x	y
0	4
2	2
4	0

Common solution: (1, 3)

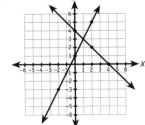

2. $y = 2x + 3$ $y = x + 1$

x	y
−1	1
0	3
1	5

x	y
0	1
2	3
4	5

Common solution: (−2, −1)

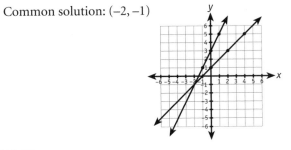

3. $y = 3x + 1$ $y = -x + 5$

x	y
−1	−2
0	1
1	4

x	y
−1	6
0	5
6	−1

Common solution: (1, 4)

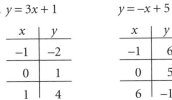

Page 113

1. a.

x	y
−4	11
−3	4
−2	−1
0	−5
2	−1
3	4
4	11

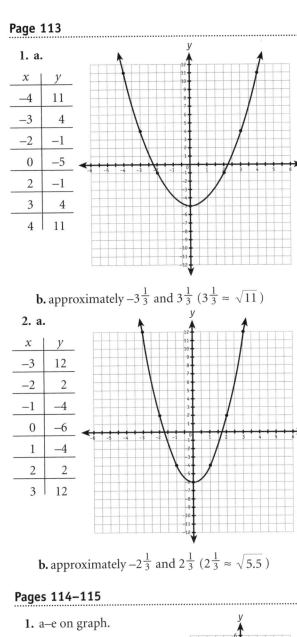

b. approximately $-3\frac{1}{3}$ and $3\frac{1}{3}$ ($3\frac{1}{3} \approx \sqrt{11}$)

2. a.

x	y
−3	12
−2	2
−1	−4
0	−6
1	−4
2	2
3	12

b. approximately $-2\frac{1}{3}$ and $2\frac{1}{3}$ ($2\frac{1}{3} \approx \sqrt{5.5}$)

Pages 114–115

1. a–e on graph.

2. Point A = (−4, 5)
Point B = (4, 1)
Point C = (2, −4)
Point D = (−2, −2)

3. a. See graph.
  b. x-intercept $= (6, 0)$
      y-intercept $= (0, -2)$
  c. slope $= \frac{1}{3}$

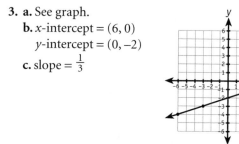

4. a. $2\left(\frac{4-0}{2-0}\right)$ **b.** $5\left(\frac{5-0}{2-1}\right)$ **c.** $0\left(\frac{3-3}{4-2}\right)$

For problem 5, Table of Values will vary.

5.

x	y
0	-4
2	0
4	4

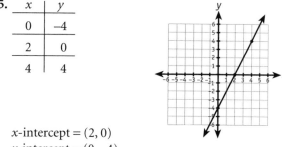

x-intercept $= (2, 0)$
y-intercept $= (0, -4)$
slope $= 2$

6. Common solution $= (3, 5)$

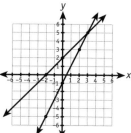

7.

x	y
-4	10
-3	3
-2	-2
0	-6
2	-2
3	3
4	10

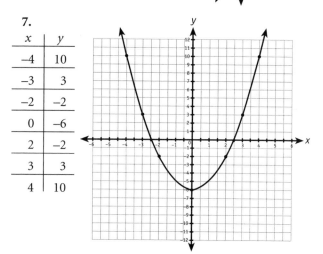

Page 116

1. a. $-3, 0, 5$ **b.** $-2, 0, 3, 7$

2. a. $8, 10, 19$ **b.** $6, 11, 23$

3. a. $-8, 4, 12$ **b.** $-5, 0, 3, 5, 6$

4. a. $17, 29, 41$ **b.** $-4, -3, 0, 4$

Pages 117–118

1. c. $s \geq -3$

2. a. $r < 1$

3. $s \geq -2$

4. $d < 4$

5.

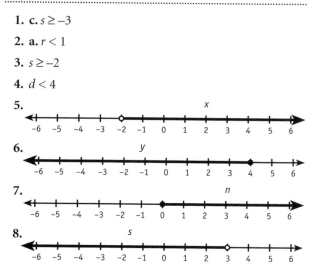

6.

7.

8.

Page 120

1. $x \geq 2, n \geq 5, z > 4$

2. $w \leq 2, x < -2, y \leq -2$

3. $n > 12, x \geq 40, c < 4$

4. $x \leq 6, n < -8, y \leq 40$

5. $x \leq -8, n > -6, x < -2$

Page 121

1. $x + x + x + 3 < 24$ or $3x + 3 < 24$; $x < 7$

2. $y + y + y + 14 + 8 < 70$ or $3y + 22 < 70$; $y < 16$

3. $l + 8 + l > 32$ or $2l + 8 > 32$; $l > 12$

4. $2d + d + 5 + 7 > 36$ or $3d + 12 > 36$; $d > 8$

Page 122

1. c. $n + 6 > 10$, $n > 4$

2. a. $3m - 8 \leq 4$, $m \leq 4$

3. a. $\$1.50n + \$35.00 \geq \$85.00$ or $\$1.50n \geq \$50.00$
   b. 34 posters ($n \geq 33\frac{1}{3}$)

4. a. $c + 26 + 17 > 75$ or $c + 43 > 75$
   b. 32 yards (The smallest value c may be is actually a little over 32 yards, a length that rounds to 32 yards.)

5. a. $54w \leq 1{,}800$
   b. 33 feet ($w \leq 33\frac{1}{3}$)

Pages 123–125

1. n is greater than -5 and n is less than 8.

2. x is greater than or equal to -1 and x is less than 11.

3. n is less than or equal to 0 or n is greater than 7.

4. p is less than -3 or p is greater than or equal to 5.

5. **a.** Yes   **b.** No   **c.** Yes

6. **a.** No   **b.** No   **c.** Yes

7. **a.** No   **b.** Yes   **c.** Yes

8. **a.** Yes   **b.** Yes   **c.** No

9. **a.** Yes   **b.** No   **c.** Yes

10. **a.** Yes   **b.** Yes   **c.** Yes

11. $-5 < x \le 4$

12. $1 \le r < 3$

13. $y < -2$ or $y \ge 2$

14. $z \le -3$ or $z > 0$

15.

16.

17. $132 \le w \le 138$

18. $0.05(2,000) \le n \le 0.1(2,000)$ or $100 \le n \le 200$

19. **a.** $102 < c + 38 + 25 < 119$ or $39 < c < 56$
    b. 40 inches

20. $t < 1$ or $t > 5$

Pages 126–127

1. x is less than 6.

2. n is greater than 9.

3. y is greater than or equal to -5.

4. s is less than or equal to 4.

5. $p > -3$ 7. $n < 2$

6. $r \le -1$ 8. $w \ge -5$

9.

10.

11. $x \le 2$

12. $n \ge 10$

13. $n + n + 5 < 45$ or $2n + 5 < 45$; $n < 20$

14. $x + 7 + x > 23$ or $2x + 7 > 23$; $x > 8$

15. x is greater than -2 and x is less than 7.

16. y is greater than or equal to 0 and y is less than 6.

17. n is less than or equal to -2 or n is greater than or equal to 5.

18. r is less than 2 or r is greater than 7.

19. $-4 < y \le 2$

20. $-3 \le n < 4$

21. **a.** $85 < BC + 31 + 36 < 94$ or $18 < BC < 27$
    b. 19 centimeters

Page 128

1. binomial   4. binomial

2. trinomial   5. trinomial

3. monomial   6. monomial

7. $-12x$ and 7; $9d$ and 6;
   $-3a^2$ and $5a$; $4y$ and $\frac{1}{2}$

8. $9a$ and $-2b$; $-6d$ and $-8e$;
   $8x^2$ and $-3y^2$; $\frac{2}{3}z^2$ and $-7z$

9. $2x^2, 7x$, and -4; $3y^2, -5y$, and -8; $8x, -4y$, and 13

Page 129

1. -4 and x, 3 and a, -5 and z

2. 1 and y^2, 1 and z^2, -1 and a^2

3. -9 and x^2y, -4 and c^2d, 7 and uv^2

4. 3 and a^2b^2, 8 and x^2y^2, -7 and ab^3

5. x and $-5x$, y^3 and $2y^3$, a^2b and $2a^2b$

6. a^2 and $-4a^2$, $2xy^2$ and xy^2, x^3 and $4x^3$

7. c^2 and $-5c^2$, $-rs^2$ and $3rs^2$, $-4x^3y$ and $2x^3y$

Pages 130–131

1. $9y, 7r, 19s, 15x$

2. $5z, 8a, 3c, 12n$

3. $5x, 7y, 11r, 0$

4. $-2x^2, 5y^2, 5z^3, 8r^3$

5. $3y + 12, 4x + 13, 7z + 3$

6. $-z + 2, 8y - 8, 16s + 4$

7. $16n^2 + n, 6x^2 + 8x, r^2 + r$

8. $-z + 9, 2x + 11, 4n + 8$

9. $3a - 5b, x + 4y, 5r - s$

10. $3z^2 + 2, 3n^2 + 3, y^2 + 4y$

11. $6a^2 - 2b, 8r^2 - 3s, 13m^2 - 2n$

12. $8m + 2n - 1, 3y^2 + 2x + 1, 4s^2 + 2s - t$

Pages 132–133

1. $3x$, $5z$, $4r$, $7x$

2. $9z$, $10b$, $8d$, $38n$

3. $2x + 2y$, $6m + 3n$, $5r - 13s$

4. $3x - 8y$, $3a + 5b$, $12c - 10d$

5. $6a - 4b$, $3x - 4y$, $7m - 2n$

6. $3x^2 + 4x - 9$, $6a^2 + 2a - 9$, $7n^2 - 2n - 4$

7. $6a - 5b$ $[(8a - 2b) - (2a + 3b)]$

8. $5x^2 - 2x + 7$ $[(12x^2 - 5x + 7) - (7x^2 - 3x)]$

9. $3n^2 - n$ $[(8n^2 + 4n + 5) - (5n^2 + 5n + 5)]$, where $5n^2 + 5n + 5$ is the sum of the three labeled sides.

Page 134

1. $15y^5$, $12x^6$, $21n^5$, $48a^7$

2. $25x^4$, $36n^8$, $16z^{12}$, $64x^6$

3. $-6x^3$, $-20c^8$, $-63b^5$, $-45s^4$

4. $-7x^3y^2$, $-5c^4d^4$, $-12a^4b^5$, $56r^3s^4$

Page 135

1. $12n^3 + 8n$, $5c^2 + 35c$, $12a^4 - 30a$

2. $6x^4 + 12x^3$, $8y^3 - 6y^2$, $9r^4 + 24r^3$

3. $3a^3b^3 - 6ab$, $5c^4d^2 + 3cd$, $x^3y^4 - 2xy$

4. $-8m^4n^3 - 4m^2n^2$, $3a^4b^2 - 18a^3b^2$, $15c^3d^3 - 10c^2d^3$

5. $12x^4 - 9x^2y + 3x^2y^2$, $16a^3 + 24a^2b^2 - 28ab^3$, $6c^4d^4 - 12c^3d^3 + 6c^2d^2$

Pages 136–137

1. $8y^2 + 16y + 6$, $8a^2 + 8a + 2$, $15z^2 + 19z + 6$

2. $8x^2 + 2xy - 3y^2$, $6a^2 + 5ab - 6b^2$, $3r^2 - 10rs - 8s^2$

3. $a^2 - 4$, $c^2 - 25$, $y^2 - 64$

4. $36 - x^2$, $81 - b^2$, $16 - m^2$

5. $10y^2$ 7. $c^2 - 36$ 9. $x^4 + 3x^2$

6. $3a^3b - 12ab^2$ 8. $64y^3$ 10. $9\pi b^3c^2$

Page 138

1. y^4, x^3, a, c

2. $3b^3$, $-3n$, $-4x^3$, $-7d^4$

3. a^2b, m^3n^2, $-4c^2d$, $-3x^3y^3$

Page 139

1. $3x^3 + 2x^2$, $4n^3 + 2n^2$, $5y^4 - 3y^2$

2. $-6n^4 + 2n^3$, $4x^4 - 2x^2$, $-2y^3 - y$

3. $2ab^4 - ab$, $3m^2 + 2mn$, $4cd - 3c$

4. $6ab^2 - 3b^3 + 4$, $3m^2n^3 + 2mn - n^2$, $3c^2d - cd + 2d$

Pages 140–141

1. binomial   17. $3a + 2b$

2. trinomial   18. $6x - 10y$

3. monomial   19. $6m - 3n$

4. binomial   20. $x^2 + x + 5$

5. $2x$ and $-5x$ 21. $20c^8$

6. n^3 and $-2n^3$ 22. $6a^3b^4$

7. cd^2 and $-cd^2$ 23. $8x^4 + 6x$

8. $-2a^2b$ and $3a^2b$ 24. $8m^3 + 2m^2n^2 - 6mn^3$

9. $11r$ 25. $6a^2 - 11a - 10$

10. $2y^2$ 26. $x^2 - x - 20$

11. $10x - 4$ 27. $35x^2 + 30x$

12. $4n^2 - 2n$ 28. $3a^3b - 6ab^2$

13. $8a + 2b$ 29. n^2

14. $3x^2 + 9x - 6$ 30. $-4a^4b^3$

15. $4s$ 31. $5c^2 + 2c$

16. $17d$ 32. $7mn - 4m + 3n$

Page 142

1. 3: 1, 3
   9: 1, 3, 9
   14: 1, 2, 7, 14

2. 24: 1, 2, 3, 4, 6, 8, 12, 24
   36: 1, 2, 3, 4, 6, 9, 12, 18, 36
   64: 1, 2, 4, 8, 16, 32, 64

3. 5, 2, 4, 5

4. 5, 7, 15, 7

5. 2, 2; 3, 3; 5, 2

6. 3, 2; 7, 2; 5, 2

Page 143

1. 1, 11; prime

2. 1, 2, 5, 10; composite

3. 1, 19; prime

4. 1, 3, 7, 21; composite

5. 1, 2, 11, 22; composite

6. 1, 29; prime

7. $2 \times 2 \times 2 \times 2$ or 2^4

8. $5 \times 2 \times 2$ or $2^2 \cdot 5$

9. $2 \times 2 \times 2 \times 2 \times 2$ or 2^5

10. $7 \times 2 \times 2 \times 2$ or $2^3 \cdot 7$

11. $5 \times 5 \times 3$ or $3 \cdot 5^2$

12. $3 \times 2 \times 2 \times 2 \times 2 \times 2$ or $2^5 \cdot 3$

Page 144

1. y, n^2, y^4, r^3

2. $5x^2, 4y^4, 6n^3, 7a^5$

3. $2x^2y^3, 5a^4b, 9c^3d^2, 8x^3y$

Page 145

1. $4\sqrt{2}, 3\sqrt{5}, 5\sqrt{3}$

2. $x\sqrt{y}, x^3\sqrt{y}, b^4\sqrt{a}$

3. $2y\sqrt{x}, 4z^3\sqrt{y}, 5r^2\sqrt{s}$

Pages 146–147

1. $2(x+3), 3(x+4), 5(y-4)$

2. $3(x^2+2x+1), 4(x^2-2x+3), 7(y^2-2y-5)$

3. $x(2x+5), x(3x-7), z(2z+11)$

4. $y(y^2-2y+4), z(z^2+6z-4), x(x^2-3x+2)$

5. $2(w^2+4), 3(3x^2+1), 4(3y^2-2)$

6. $5x(x+2), 6x(x+4), 7z(z-3)$

7. $6z(2z-3), 2y(2y+3), 5z(2z-3)$

8. $3y^2(y+5), 5x^2(x-5), 4s^2(s+4)$

9. $4x^2(2x^2+3), 3y^2(3y^2+2), 5z^2(3z^2+4)$

10. $2ab(b+2a), 3xy(2x+1), 2rs(2r+3s)$

Pages 148–149

1. $x^2+2, 4r^2+3$

2. $2y+1, n+3$

3. $x-2, 2-ab$

4. $3, 4$

5. x, n

6. $3r, 7v$

7. length $= 4x$

8. width $= 2+b$

9. height $= 2z^2+z+4$

10. height $= 8ab$

Pages 150–151

1. $(5+s)(5-s), (6+t)(6-t)$

2. $(8+n)(8-n), (4+x)(4-x)$

3. $(r+7)(r-7), (y+9)(y-9)$

4. $x(x+5)(x-5)$
$y(y+6)(y-6)$
$n(n+4)(n-4)$

5. $4s(s+3)(s-3)$
$8x(x+2)(x-2)$
$2y(y+4)(y-4)$

6. $d+9, t-8$

7. $5-r, 4n(n+3)$ or $4n^2+12n$

8. length $= 7+x$

9. width $= a-8$

10. height $= 4n(n-2)$ or $4n^2-8n$

Pages 152–153

1. $1, 3, 9$

2. $1, 7$

3. $1, 2, 3, 4, 6, 8, 12, 24$

4. $4, 2$

5. $3, 2$

6. $7, 3$

7. 7×2

8. $2 \times 2 \times 2 \times 2$ or 2^4

9. $5 \times 5 \times 2$ or $2 \cdot 5^2$

10. $4x^2$

11. $6ab$

12. $5c^2d^3$

13. $3\sqrt{2}$

14. $5x\sqrt{y}$

15. $7d^3\sqrt{c}$

16. $4(x+1)$

17. $3(y-5)$

18. $6(y+4)$

19. $x(3x+7)$

20. $n(2n^2-9)$

21. $y^2(5y^2+2)$

22. $3x(x+4)$

23. $5n^2(n-3)$

24. $2xy(4x+1)$

25. $x+1$

26. n^2+2

27. $2-mn$

28. 4

29. y

30. $3b$

31. $(6+n)(6-n)$

32. $(x+5)(x-5)$

33. $x-7$

34. $3(n+3)$

35. length $= 3ab+b$

36. width $= x-6$

Pages 154–159, Posttest A

1. $-2 + -4 = -6$
2. $5 + (-4) = 1$
3. 15
4. 3
5. -32
6. $\frac{5}{9}$
7. 81
8. -64
9. 7^6
10. 10^4
11. 9
12. $\frac{6}{5}$
13. $\frac{1}{4}(x - 6)$
14. $14(20 - 6)$
15. $-12 [-12(\frac{2}{2})]$
16. $25°C [\frac{5}{9}(45)]$
17. a. $n - 4 = 10$

18. a. $n = -6$ b. $x = 37$
19. a. $p = 9$ b. $r = 66$
20. a. $x = 9$ b. $a = 15$
21. a. $n = 9$ b. $x = 6$
22. a. $s = 13$ b. $y = 12$
23. length = 57 yards, width = 38 yards
24. $144 \left(\frac{2}{5} = \frac{n}{360}\right)$
25. $x = \pm 6$
26.

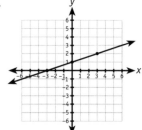

27. slope $= -\frac{1}{5}$
28.

x	y
0	−4
2	2
3	5

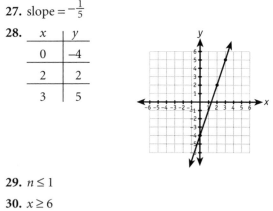

29. $n \le 1$
30. $x \ge 6$
31. $r < 7$
32.

Values of s

33. $2x^3$ and $-3x^3$

34. $n^2 - 2$
35. $4x^2 - 9x$
36. $8d^3 - 4d^2 + 4d$
37. $7n^3 + 5n$
38. $6\sqrt{2}$
39. $7a^3\sqrt{b}$
40. $3(3x^2 + 4)$
41. $5n^2(4n + 3)$
42. $2x^2 + 1$
43. 9

Pages 161–166, Posttest B

1. e. $4 + (-3)$
2. c. $-24°F$
3. b. $16 [21 + (-5)]$
4. e. 16
5. a. 6^5
6. b. $\frac{6}{5}$
7. c. $4(n - 7)$
8. d. $40 [(5)(8)]$
9. a. $10°C [\frac{5}{9}(18)]$
10. c. $n - 8 = 19$
11. e. Add +8 to each side of the equation.
12. d. Subtract 9 from each side of the equation.
13. a. $x = 18$
14. c. $n = 4$
15. d. $x + (x + \$40) + 2x = \$2,620$
16. c. 55
17. c. two
18. d. $\frac{n}{18} = \frac{3}{5}$
19. e. $(4, 0)$
20. b. -1
21. c. $(1, 2)$ (If $x = 1$, then $y = -1$, not 2.)
22. e. 6
23. a. $3x + 6 < 35$
24. b. $-4 < x \le 4$
25. a. $3n^2 - 3n$
26. e. $13x^2 + 8$
27. d. $x^4 + 4x^3$
28. a. $7x^3\sqrt{y}$
29. d. $8n^2(2n^2 - 3)$
30. e. $r - 3$

Page 169

1. $-79°F$
2. $-27°F$
3. $-53°F$
4. $-5°F$
5. difference $= 45° [-15° - (-60°)]$
6. difference $= 40° [15° - (-25°)]$
7. 20 miles per hour
8. $-5°F$

Page 171

1. 600 miles (50×12)
2. 55 miles per hour ($440 \div 8$)
3. 7 hours ($350 \div 50$)
4. 288 miles (48×6)
5. 52 miles per hour ($364 \div 7$)
6. 6 hours ($324 \div 54$)

Page 173

1. $30 (interest = $200 × $\frac{5}{100}$ × 3)

2. $721.50 [total = $650 + ($650 × 0.055 × 2)]

3. $95 (interest = $475 × $\frac{1}{10}$ × 2)

4. $1,330 [total = $1,000 + ($1,000 × $\frac{11}{100}$ × 3)]

5. $71.88 (interest = $575 × 0.125 × 1)

6. $453.75 [total = $375 + ($375 × 0.0525 × 4)]

Page 175

1. $67.50 (interest = $900 × $\frac{5}{100}$ × $\frac{3}{2}$)

2. $1,320 [total = $1,000 + ($1,000 × $\frac{12}{100}$ × $\frac{8}{3}$)]

3. $85 (interest = $850 × $\frac{6}{100}$ × $\frac{5}{3}$)

4. $1,120 [total = $1,000 + ($1,000 × $\frac{16}{100}$ × $\frac{3}{4}$)]

5. $112.50 (interest = $750 × $\frac{18}{100}$ × $\frac{10}{12}$)

6. $8,102.50 [total = $7,000 + ($7,000 × 0.105 × 1.5) or $7,000 + ($7,000 × $\frac{10.5}{100}$ × $\frac{3}{2}$)]

Page 177

1. $\frac{1}{2}$ hour = 30 minutes

($1x + 1x = 1$ or $2x = 1$)

2. $1\frac{1}{3}$ hours = 1 hr 20 min

($\frac{1}{2}x + \frac{1}{4}x = 1$)

3. $1\frac{5}{7}$ hours ≈ 1 hr 43 min

($\frac{1}{4}x + \frac{1}{3}x = 1$)

4. $2\frac{2}{9}$ hours ≈ 2 hr 14 min

($\frac{1}{5}x + \frac{1}{4}x = 1$)

5. $1\frac{7}{8}$ hours ≈ 1 hr 53 min

($\frac{1}{3}x + \frac{1}{5}x = 1$)

Page 179

1. 10 pounds of nuts

[$6x + 3(5) = 5(x + 5)$ or $6x + 15 = 5x + 25$]

2. 4 pounds of dried fruit mix

[$3x + 4.5(8) = 4(x + 8)$ or $3x + 36 = 4x + 32$]

3. 1 pound of the $2.00 per pound chocolates

[$2x + 1.6(3) = 1.7(x + 3)$ or $2x + 4.8 = 1.7x + 5.1$]

Pages 180–181

1. 32°F

2. 77°F

3. 356°F

4. 100.4°F

5. 68°F

6.

°C	°F
0	32
10	50
30	86

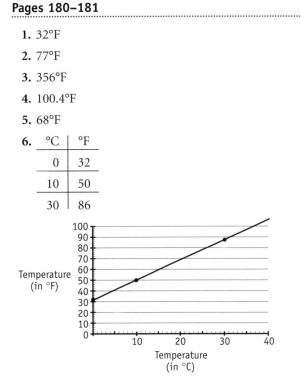

7. about 80°F (exactly 82.4°F)

8. about 37°C (exactly 37.7°C)

9. about 2 (exactly $\frac{9}{5}$)

slope = change in F temperature ÷ change in C temperature. Notice that the scales of axes are different.

Page 183

1. ∠EBC = 10°; ∠DBE = 30°; ∠ABD = 50°

($x + 3x + 5x = 90°$)

2. ∠D = 30°; ∠E = 90°; ∠F = 60°

($n + 2n + 3n = 180°$)

3. ∠EBC = 6°; ∠DBE = 30°; ∠ABD = 54°

($x + 5x + 9x = 90°$, where $x = $∠EBC)

4. ∠A = 18°; ∠B = 72°; ∠C = 90°

($n + 4n + 5n = 180°$, where $n = $∠A)

Page 185

1. 10 feet ($c^2 = 6^2 + 8^2$)

2. 9 inches ($b^2 = 15^2 - 12^2$)

3. 15 feet ($b^2 = 17^2 - 8^2$)

4. 13 yards ($c^2 = 12^2 + 5^2$)

5. about 8.6 feet ($c^2 = 5^2 + 7^2$)

6. about 12 miles ($c^2 = 10^2 + 7^2$)

Page 187

1. points A and B: 5 units ($c^2 = 3^2 + 4^2$)

2. points A and C: 10 units ($c^2 = 8^2 + 6^2$)

3. points A and B: 5 units

4. points B and C: 12 units

5. points A and C: 13 units [$c^2 = 5^2 + 12^2$]

6. points A and B: $6\frac{1}{2}$ units ($c^2 = 6^2 + \left(\frac{5}{2}\right)^2$

$$= 36 + \frac{25}{4}$$
$$= \frac{144}{4} + \frac{25}{4}$$
$$= \frac{169}{4}$$
$$c = \sqrt{\frac{169}{4}} = \frac{13}{2}$$
$$= 6\frac{1}{2}$$

Page 189

1. $4, 0, 8.5, \frac{1}{2}, 9$

2. $2, 9, 7$

3. $x = -5$ and $x = 5$
   $n = -3$ and $n = 3$
   $r = -10$ and $r = 10$
   $w = 0$
   $z = -100$ and $z = 100$

4. $x = -1$ and $x = 5$
   $n = 4$ and $n = 6$
   $r = -8$ and $r = 22$

5.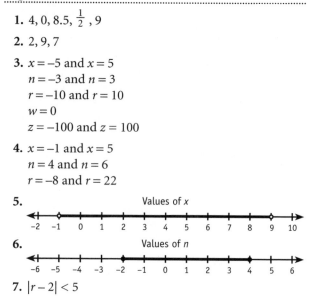

6.

7. $|r - 2| < 5$

Using a Calculator

A calculator is a valuable math tool. You'll use it mainly to add, subtract, multiply, and divide quickly and accurately. The calculator pictured at the right is similar to one you've seen or may be using.

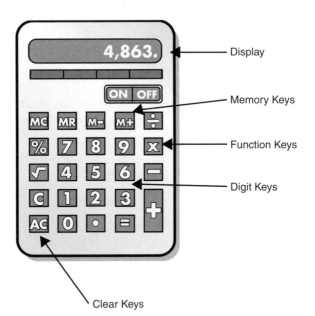

Display

Memory Keys

Function Keys

Digit Keys

Clear Keys

To enter a number, press one digit key at a time. On this display, the number 4,863 is entered. Notice the following features.

- A decimal point is displayed to the right of the ones digit.
- The calculator does not have a comma (,) key or a dollar sign ($) key.
- Pressing the clear key $\boxed{C}$ erases the display. You should press $\boxed{C}$ each time you begin a new problem or when you've made a mistake.

EXAMPLE 1 Multiply 3.14 (π) by 9.

STEP 1 Press $\boxed{C}$ to clear the display.

STEP 2 Press the digit keys and decimal point key, multiply key $\boxed{\times}$, and equals key $\boxed{=}$.

Press Keys	Display Shows
$\boxed{C}\ \boxed{3}\ \boxed{.}\ \boxed{1}\ \boxed{4}\ \boxed{\times}\ \boxed{9}\ \boxed{=}$	28.26

ANSWER: 28.26

EXAMPLE 2 Divide $23.50 by 5.

STEP 1 Press $\boxed{C}$ to clear the display.

STEP 2 Press the digit keys and decimal point key, division key $\boxed{\div}$, and equals key $\boxed{=}$.

Press Keys	Display Shows
$\boxed{C}\ \boxed{2}\ \boxed{3}\ \boxed{.}\ \boxed{5}\ \boxed{0}\ \boxed{\div}\ \boxed{5}\ \boxed{=}$	4.7

ANSWER: $4.70

Press the decimal point key to separate dollars from cents.

Calculators do not show zeros at the right end of a decimal fraction.

Using Estimation and Mental Math

To **estimate** is to compute an approximate answer. Estimating is often done as mental math—math done in your head. One way to estimate is to use rounded numbers that are easy to work with. You can estimate to

- discover about what an exact answer should be
- help pick a correct answer from among multiple choices, such as on multiple-choice questions found on many tests
- quickly check an answer obtained when using a calculator—making sure that you worked the problem correctly *and* that you did not make a keying error

In *Number Power Algebra* there are questions for which you are asked to estimate an answer. On these questions, you do not need to calculate an exact answer. Here are some general guidelines for estimating using rounded numbers.

Whole Numbers

To estimate with whole numbers, replace each number by a rounded number that contains one or more ending zeros.

<u>EXAMPLE 1</u> Multiply: 189×43
Estimate: $200 \times 40 = \textbf{8,000}$

> Round 189 to the nearest hundred; round 43 to the nearest 10.

Mixed Numbers

Round mixed numbers by replacing each mixed number with the nearest whole number.

<u>EXAMPLE 2</u> Add: $6\frac{7}{8} + 3\frac{1}{4} + 2\frac{1}{2}$
Estimate: $7 + 3 + 3 = \textbf{13}$

> Round $6\frac{7}{8}$ to 7; $3\frac{1}{4}$ to 3; and $2\frac{1}{2}$ to 3.

Decimals

Round decimals by replacing each decimal with the nearest whole number.

<u>EXAMPLE 3</u> Multiply: 7.16×2.9
Estimate: $7 \times 3 = \textbf{21}$

> Round 7.16 to 7; round 2.9 to 3.

Formulas

To estimate with formulas, round whole numbers, mixed numbers, and decimals as discussed above. Round π to 3 for a quick calculation.

<u>EXAMPLE 4</u> Multiply: $\pi \times 5\frac{1}{8} \times 5\frac{1}{8}$
Estimate: $3 \times 5 \times 5 = \textbf{75}$

> Round π to 3; round each $5\frac{1}{8}$ to 5.

Formulas

PERIMETER

Figure	Name	Formula	Meaning
	Rectangle	$P = 2l + 2w$	l = length w = width
	Square	$P = 4s$	s = side
	Triangle	$P = s_1 + s_2 + s_3$	s_1 = side 1 s_2 = side 2 s_3 = side 3
	Polygon (n sides)	$P = s_1 + s_2 + \ldots s_n$	s_1 = side 1, and so on
	Circle (circumference)	$C = \pi d$ or $C = 2\pi r$	$\pi \approx 3.14$ or $\frac{22}{7}$ d = diameter r = radius

AREA

Figure	Name	Formula	Meaning
	Rectangle	$A = lw$	l = length w = width
	Square	$A = s^2$	s = side
	Parallelogram	$A = bh$	b = base h = height
	Triangle	$A = \frac{1}{2}bh$	b = base h = height
	Circle	$A = \pi r^2$	$\pi \approx 3.14$ or $\frac{22}{7}$ r = radius

VOLUME

Figure	Name	Formula	Meaning
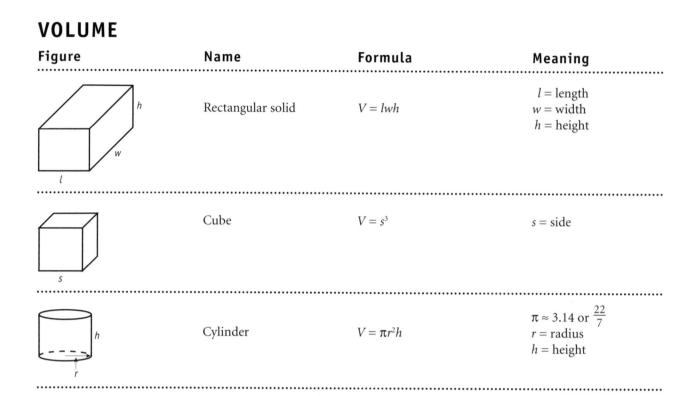	Rectangular solid	$V = lwh$	l = length w = width h = height
	Cube	$V = s^3$	s = side
	Cylinder	$V = \pi r^2 h$	$\pi \approx 3.14$ or $\frac{22}{7}$ r = radius h = height

OTHER FORMULAS

Name	Formula or Example	Meaning
Difference of squares	$a^2 - b^2 = (a + b)(a - b)$	factors of a difference of squares
Distance formula	$d = rt$	distance = rate × time
Dividing like bases	$\dfrac{x^a}{x^b} = x^{a-b}$	dividing like bases
Interest	$i = prt$	i = interest, p = principal, r = rate (%), t = time (yr)
Linear equation	$y = 4x - 1$	two-variable equation in which each variable has an exponent of 1
Multiplying like bases	$x^a x^b = x^{a+b}$	multiplying like bases
Pythagorean theorem	$c^2 = a^2 + b^2$	c = hypotenuse; a and b are the two shorter sides of the right triangle
Quadratic equation	$y = 2x^2 + 3$	two-variable equation in which one variable is squared
Slope	$\dfrac{\text{Change in } y \text{ values}}{\text{Change in } x \text{ values}}$	slope of a line on a coordinate grid
Temperature	$°F = \dfrac{9}{5}°C + 32°$	°F = degrees Fahrenheit °C = degrees Celsius

Glossary

A

absolute value   The positive value of a number or expression regardless of its sign. For example, $|5| = 5$ and $|{-5}| = 5$

algebraic expression   Two or more numbers or variables combined by addition, subtraction, multiplication, or division

***and* inequality**   An inequality, such as $2 \leq x < 6$, in which the allowed values of the variable run between two numbers

area (*A*)   An amount of surface. Area is measured in square units.

$$\begin{aligned} \text{area} &= l \times w \\ &= 8 \times 6 \\ &= 48 \text{ square yards} \end{aligned}$$

6 yd

8 yd

axis   One of the perpendicular sides of a graph along which numbers, data values, or labels are written. Plural is *axes*.

vertical axis

horizontal axis

B

base   The number being multiplied in a power. In 2^3, 2 is the base; 3 is the exponent.

binomial   A polynomial of two terms. Examples: $x + 4$, $3y^2 - 1$, $r^3 + 5$

C

circle   A plane (flat) figure, in which each point is an equal distance from the center

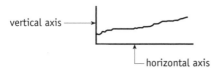

center

circle

circumference   The distance around a circle
$C = \pi \times \text{diameter}$ ($\pi \approx 3.14$)

circumference

coefficient   A number that multiplies a variable. In the term $3y$, 3 is the coefficient.

comparison symbols   Symbols used to compare one number with another

Symbol	Meaning	Example	Meaning
$<$	is less than	$n < 7$	n is less than 7
$>$	is greater than	$x > 6$	x is greater than 6
$\leq$	is less than or equal to	$y \leq 2$	y is less than or equal to 2
$\geq$	is greater than or equal to	$s \geq -3$	s is greater than or equal to -3

composite number   A number, such as 12, that has more than two factors

coordinates   A set of two numbers that identifies a point on a coordinate grid

coordinate axis   A number line that forms the horizontal axis or vertical axis of a coordinate grid. See *horizontal axis* and *vertical axis*.

coordinate grid   A grid on which points are identified by an ordered pair of numbers called coordinates

cross products   In a proportion, the product of the numerator of one ratio times the denominator of the other. In a true proportion, cross products are equal.

Proportion	Equal Cross Products
$\frac{2}{3} = \frac{6}{9}$	$2 \times 9 = 3 \times 6$
	$18 = 18$

cube   A 3-dimensional shape that contains six square faces. At each vertex, all sides meet at right angles.

face

cube

cylinder   A 3-dimensional shape that has both a circular base and a circular top

cylinder

D

degrees   The measure (size) of an angle. A circle contains 360 degrees (360°). One-fourth of a circle contains 90°.

diagonal   A line segment running between two nonconsecutive vertices of a polygon. A diagonal divides a square or rectangle into two equal right triangles.

diagonal

diameter   A line segment, passing through the center, from one side of a circle to the other. The length of the diameter is the distance across the circle.

diameter

distance formula   The formula $d = rt$ relates distance, rate, and time. In words, *distance* equals *rate* times *time*. By rearranging the variables, you can write the rate formula $r = \frac{d}{t}$ or the time formula $t = \frac{r}{d}$.

E

equation   A statement that two quantities are equal (have equal value or measure)

equivalent expressions   Expressions that have the same value. The expression $2x + 4x + 5$ is equivalent to the expression $6x + 9 - 4$. Both expressions simplify to $6x + 5$.

exponent   A number that tells how many times the base (of a power) is written in the product. For example, in 5^2, 2 is the exponent.

estimate   To find an approximate answer. For example, by calculating with rounded numbers

F

factor   A number that divides evenly into another number. Numbers and variables can be factors of an algebraic experession. Example: 1, 2, 4, and 8 are factors of 8.

formula   A mathematical rule that tells the relationship between quantities. For example, the formula relating the area of a circle to its radius is $A = \pi r^2$.

H

horizontal   Running right and left

horizontal axis   On a coordinate grid, the axis running from left to right

horizontal axis

hypotenuse   In a right triangle, the side opposite the right angle

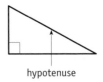

hypotenuse

I

inequality   A statement that two numbers are not equal. Examples: $n \geq 6$ states that n is greater than or equal to 6; $-2 < x \leq 3$ states that x is greater than -2 and less than or equal to 3.

intercept   A point where a graphed line crosses a coordinate axis. The x-intercept is the point where the line crosses the x-axis. The y-intercept is the point where the line crosses the y-axis.

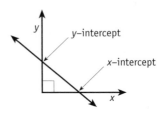

inverse operation   An operation that "undoes" a result: subtraction is the inverse operation of addition; division is the inverse operation of multiplication.

L

like terms   Terms that have the same variable raised to the same power: 4 and 7 are like terms (no variable); $2r$ and $5r$ are like terms; $3x^2$ and $-2x^2$ are like terms.

linear equation   An equation such as $y = 2x$ that, when graphed, forms a straight line

M

monomial   A polynomial of one term Examples: x, y^3, $5r^2$

multistep equation   An equation such as $2x + 5 = 9$ that contains more than one operation

N

negative exponent   An exponent that is less than 0. A negative exponent stands for the number found by inverting a power.

negative number   A number less than zero. A negative number is written with a negative sign. Examples: -6, -2.5, -1

number line A line used to represent positive numbers, negative numbers, and 0. A number line can be written horizontally (left to right) or vertically (up and down).

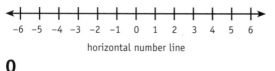

horizontal number line

O

or **inequality** An inequality, such as $r < 4$ *or* $r \geq 9$, in which the allowed values of the variable are graphed as separate sections of a number line

ordered pair A pair of numbers that is used to identify a point on a coordinate grid

P

perfect square A number whose square root is a whole number. Example: 64 is a perfect square; $\sqrt{64} = 8$.

perimeter The distance around a flat (plane) figure. The symbol for perimeter is P.

perpendicular lines Lines that meet (or cross) at a right angle (90°)

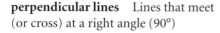

perpendicular lines

pi (π) The ratio of the circumference of a circle to its diameter. Pi is approximately equal to 3.14 or $\frac{22}{7}$.

polynomial An algebraic expression that contains one or more terms. Monomials, binomials, and trinomials are all polynomials.

positive number A number greater than 0. Positive numbers may be written with a positive sign (+) or no sign at all.

power The product of a number multiplied by itself one or more times. Example: $5^2 = 5 \times 5 = 25$

prime factorization form Writing a number as a product of prime factors
Example: $36 = 2 \times 2 \times 3 \times 3 = 2^2 3^2$

prime number Any number greater than 1, such as 7, that has only two factors: itself and the number 1

proportion Two equal ratios. A proportion can be written with colons or as equal fractions.

Written with colons: $2:3 = 6:9$
Written as equal fractions: $\frac{2}{3} = \frac{6}{9}$

Pythagorean theorem A theorem that states: In a right triangle, the square of the hypotenuse is equal to the sum of the squares of the two shorter sides.

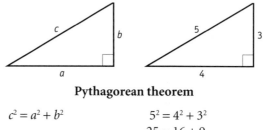

Pythagorean theorem

$$c^2 = a^2 + b^2 \qquad\qquad 5^2 = 4^2 + 3^2$$
$$25 = 16 + 9$$
$$25 = 25$$

Q

quadratic equation An equation such as $y = x^2$ that contains the square of a variable

R

radius A line segment from the center of a circle to any point on the circumference of the circle

radius

ratio A comparison of two numbers. For example, "4 to 3." A ratio can be written as a fraction or with a colon.

4 to 3 is written $\frac{4}{3}$ or 4:3

reciprocal A number formed by inverting a fraction or whole number: $\frac{3}{2}$ is the reciprocal of $\frac{2}{3}$; $\frac{1}{4}$ is the reciprocal of 4.

rectangle A four-sided polygon with two pairs of parallel sides and four right angles

rectangle

rectangular solid (prism) A 3-dimensional figure in which each face is either a rectangle or a square. Opposite faces are congruent.

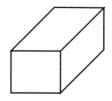

rectangular solid

reflex angle An angle that measures greater than 180° but less than 360°

reflex angle

right angle An angle that measures exactly 90°. A right angle is often called a "corner angle."

right angle

right triangle A triangle that contains a right angle

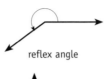
right triangle

S

scientific notation Writing a number as a product of two factors: a number between 1 and 10 *and* a power of 10. Examples: $5{,}300 = 5.3 \times 10^3$, $0.075 = 7.5 \times 10^{-2}$

signed numbers The set of all negative numbers and positive numbers

simple interest Interest earned (or paid) on an unchanging principal. The simple interest formula is $i = prt$.

slope The rise or fall of a graphed line. The slope is defined as the change in y values divided by the change in x values.

Positive slope: The line rises (goes up) from left to right.
Negative slope: The line falls (goes down) from left to right.
Zero slope: A horizontal line that neither rises nor falls
Undefined slope: A vertical line

square A polygon with four equal sides, two pairs of parallel sides, and four right angles

square

square of a number The product of a number multiplied by itself. Example: $8^2 = 8 \times 8 = 64$

square root ($\sqrt{\ }$) One of two equal factors of a number. Example: $\sqrt{9} = 3$

system of equations Two or more linear equations that have a common solution. Example: $y = 2x + 3$ and $y = x + 2$ have the common solution $x = -1$ and $y = 1$.

T

term A number standing alone or a variable multiplied by a coefficient. Examples: 5, $3x$, $2y^2$. Terms are connected by + and − signs.

triangle A polygon having three sides and three angles

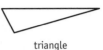
triangle

trinomial A polynomial of three terms Examples: $x^2 - 3x + 1$, $2y^3 + y + 4$

U

unknown A letter that stands for a number. Also called a *variable*. For example, in $x + 4 = 7$, x is an unknown standing for the number 3.

V

variable A letter that stands for a number. Also called an *unknown*

vertical axis On a coordinate grid, the axis running up and down

vertical axis

volume The amount of space an object takes up. Volume is measured in cubic units.

$$\begin{aligned}\text{volume} &= l \times w \times h \\ &= 5 \times 4 \times 3 \\ &= 60 \text{ cubic feet}\end{aligned}$$

3 ft
4 ft
5 ft

W

whole numbers The set of counting numbers (1, 2, 3, . . .) and the number 0

wind chill A temperature reading that takes into account the chilling effect of the wind. Wind chill temperature is lower than thermometer temperature.

work problem A problem that involves finding the rate at which two or more people working together can finish a job

Index